AF361769

# Precision Machining

# Precision Machining

Editor

**Angelos P. Markopoulos**

MDPI • Basel • Beijing • Wuhan • Barcelona • Belgrade • Manchester • Tokyo • Cluj • Tianjin

*Editor*
Angelos P. Markopoulos
School of Mechanical Engineering
National Technical University
of Athens
Athens
Greece

*Editorial Office*
MDPI
St. Alban-Anlage 66
4052 Basel, Switzerland

This is a reprint of articles from the Special Issue published online in the open access journal *Machines* (ISSN 2075-1702) (available at: www.mdpi.com/journal/machines/special_issues/precision_machining).

For citation purposes, cite each article independently as indicated on the article page online and as indicated below:

LastName, A.A.; LastName, B.B.; LastName, C.C. Article Title. *Journal Name* **Year**, *Volume Number*, Page Range.

ISBN 978-3-0365-2837-3 (Hbk)
ISBN 978-3-0365-2836-6 (PDF)

# Contents

# About the Editor

**Angelos P. Markopoulos**

Angelos P. Markopoulos is an Associate Professor at the School of Mechanical Engineering, National Technical University of Athens, Greece. His research includes topics such as precision and ultraprecision conventional and non-conventional machining processes with a special interest in advanced manufacturing and Industry 4.0. Furthermore, he is an expert in manufacturing technology modeling and simulation, including the finite element method, artificial intelligence, and molecular dynamics. He is the author of more than 140 papers in journals, conferences, and book chapters on the above-mentioned areas and a member of the editorial boards of several international journals.

# Preface to "Precision Machining"

Material removal processes, conventional and non-conventional ones, are considered exceptionally important manufacturing methods that are used for the production of mechanical components. A key feature of these processes is their ability to produce final products with high accuracy and of high quality. Conventional and non-conventional machining, as well as abrasive processes, are vital for the production of high-quality components from many different materials categories. Automotive, aerospace and medical industries are only some of the sectors that machined components of high dimensional accuracy, exceptional properties, complex sizes and usually from difficult-to-machine materials, are employed. The research in the refinement of machining or the introduction of new features is ongoing and fast-growing. Precision machining on large scale components but in the micro- and nano-regime as well, concentrate the interest of the researchers. In the success of the research there are other aspects that need to be considered as the machine tools design and control, cutting tools, metrology and quality control, manufacturing systems and automation and of course modeling and simulation with various methods such as finite elements method, molecular dynamics and soft computing.

This book aims at presenting recent advances and technologies in the aforementioned fields and indicate the future trends for precision machining. More specifically, a work on laser machining that allows the machining of complex geometries with high precision surface quality and productivity in a wide range of materials is presented in [1]. A method is proposed that helps the machine operator to select the optimal process parameters. An extensive experimental work is carried out, where the average output power, the repetition rate and the scanning speed of laser engraving on a steel plate at various combinations are studied and the effectiveness of the process through the removed material layer thickness and the material removal rate is assessed. In the next study [2], the experimental analysis of profile end-milling and more specifically the effect of cutter runout on cutter vibration and, by extension, on how this affects the chip removal and, thereby, the workpiece topomorphy, is presented. Various cutting conditions such as the cutting speed, feed rate, and the axial depth of cut are considered and the effect of cutter runout is evaluated. The third work [3], is dedicated to experimental studies on drilling with solid carbide tools workpieces of Al7075. The effect of cutting speed, feed rate, diameter of the tool on the thrust force ($Fz$) and the cutting torque ($Mz$) are evaluated and analyzed through response surface methodology (RSM) and analysis of variance (ANOVA). The diameter of cut and the feed rate were found to be the factors of high significance, while cutting speed did not considerably affect the $Fz$ and $Mz$ in the experiments that were performed. In the next interesting work [4], a knurled interference fit, i.e., a machine part connection made by a plastic joining, is examined. The combination of the friction and form fit, which are responsible for torque transmission, results in a higher power density than conventional connections. Experimental investigations on the shaft chamfer angle (100Cr6) and hub-diameter-ratio (AlSi1MgMn) are performed and analytical approaches are developed for calculating the joining force and maximal torque capacity. The presented calculation approach is an accurate tool for the assessment of machine designs of the knurled interference fit and helps to save from time-extensive tests. The work of Zhang et al. [5] pertains to annular polishing technology, which is an important optical machining method for achieving a high-precision mirror surface on silicon carbide and aims on avoiding the over-polishing of the specimen edge and thus prevent the deterioration of the surface quality. At first an analytical investigation of the the kinematic coupling of multiple relative motions in the annular polishing process is considered. Subsequently, an analytical model that addresses the principle of material removal at specimen edge based on the Preston equation and the rigid body contact model is derived. Then finite element simulations and experiments, involving annular polishing of silicon carbide (SiC), are performed, which jointly exhibit agreement with the derived analytical material removal model.

Felhő and Kundrák [6] study the effect of increasing feed per tooth on the topography of the surface in fly-cutting and in multi-point face milling. The study takes into account the axial run-out of the inserts. Theoretical roughness values were modelled, the real values were tested in experiments and in both cases the impact of the run-out of the cutting edges and the change of the chip cross-section were also taken into account. Based on the performed experiments it can be stated that the accuracy of the introduced roughness prediction method increases with the increase in feed, and therefore the application of the method in the case of high-feed milling is particularly effective. The next work [7] is an in-depth analysis of cutting procedure, which is a topic of particular interest in manufacturing efficiency; in large-scale production, the effective use of production capacities and the revenue-increasing capacity of production are key conditions of competitiveness. That is why the analysis of time and material removal rate, which are in close relation to production, are important in planning a machining procedure. Thus, three procedures applied in hard cutting are compared on the basis of these parameters and a new parameter, the practical parameter of material removal rate, is introduced. The results can give some useful indications about machining procedure selection. The following study [8] is focused on the design and manufacturing of medical implants. The methodology of designing and machining the femoral component of total knee replacement using a 3-axis Computer Numerical Control (CNC) machine is presented, and then, the results of the machining process, as well as the evaluation of implant surface quality, are discussed in detail. Analysis of the results indicated the appropriate process conditions for each part of the implant surface and led to the determination of optimum machining strategy for the finishing stage.

The double row tapered roller bearings and the manufacturing methods for producing the windows found in bearing cages are the subject of [9]. On the dimensional precision of windows, the clearance between punches and die, the work stroke length and the workpiece thickness could are crucial factors. To evaluate their influence, experimental research was undertaken, considering the height and the length of the cage window and the distance between the contact elements of the cage. The next study [10] deals with the numerical and experimental characterization of the structural behavior of a railroad switch machine. The results have been validated by means of an ad-hoc designed experimental apparatus. The last work [11] deals with a modern design approach via finite elements in the definition of the main structural elements (rotary table and working unit) of an innovative family of transfer machine tools. Using the concepts of green design and manufacture, as well as sustainable development thinking, the advantages derived from their application in this specific field is highlighted. The design is conceived in a modular way, so that the final solution can cover transfers from 4 to 15 working stations. The loading input forces for the analyses have been evaluated experimentally via drilling operations carried out on a three-axis CNC unit. The definition of the design force made it possible to accurately assess both the rotary table and the working units installed in the machine.

The guest editor would like to thank the authors for their valuable high-quality work submitted, the reviewers for their efforts and time spent in order to improve the submissions, and the publisher for their excellent work and cooperation.

**References**


1. Nikolidakis, E.; Choreftakis, I.; Antoniadis, A. Experimental Investigation of Stainless Steel SAE304 Laser Engraving Cut-ting Conditions. Machines 2018, 6, 40.

2. David, C.; Sagris, D.; Stergianni, E.; Tsiafis, C.; Tsiafis, I. Experimental Analysis of the Effect of Vibration Phenomena on Workpiece Topomorphy Due to Cutter Runout in End-Milling Process. Machines 2018, 6, 27.

3. Kyratsis, P.; Markopoulos, A.P.; Efkolidis, N.; Maliagkas, V.; Kakoulis, K. Prediction of Thrust Force and Cutting Torque in Drilling Based on the Response Surface Methodology. Machines 2018, 6, 24.

4. Suchý, L.; Leidich, E.; Gerstmann, T.; Awiszus, B. Influence of Hub Parameters on Joining Forces and Torque Transmission Output of Plastically-Joined Shaft-Hub-Connections with a Knurled Contact Surface. Machines 2018, 6, 16.

5. Zhang, J.; Han, L.; Liu, H.; Shi, Y.; Yan, Y.; Sun, T. Theoretical and Experimental Studies of Over-Polishing of Silicon Carbide in Annular Polishing. Machines 2018, 6, 15.

6. Felhő, C.; Kundrák, J. Effects of Setting Errors (Insert Run-Outs) on Surface Roughness in Face Milling When Using Circular Inserts. Machines 2018, 6, 14.

7. Kundrak, J.; Molnar, V.; Deszpoth, I. Comparative Analysis of Machining Procedures. Machines 2018, 6, 13.    8. Markopoulos, A.P.; Galanis, N.I.; Karkalos, N.E.; Manolakos, D.E. Precision CNC Machining of Femoral Component of Knee Implant: A Case Study. Machines 2018, 6, 10.

9. Rîpanu, M.; Nagîţ, G.; Slătineanu, L.; Dodun, O. The Dimensional Precision of Forming Windows in Bearing Cages. Machines 2018, 6, 9.

10. Croccolo, D.; De Agostinis, M.; Fini, S.; Olmi, G.; Robusto, F. Numerical and Experimental Characterization of a Railroad Switch Machine. Machines 2018, 6, 6.

11. Croccolo, D.; Cavalli, O.; De Agostinis, M.; Fini, S.; Olmi, G.; Robusto, F.; Vincenzi, N. A Methodology for the Lightweight Design of Modern Transfer Machine Tools. Machines 2018, 6, 2.



**Angelos P. Markopoulos**

*Editor*

*Article*

# Experimental Investigation of Stainless Steel SAE304 Laser Engraving Cutting Conditions

**Evangelos Nikolidakis [1], Ioannis Choreftakis [1] and Aristomenis Antoniadis [1,*]**

Technical University of Crete, Department of Production Engineering & Management, Micromachining & Manufacturing Modeling Lab., University Campus Kounoupidiana, 73100 Chania, Greece; krhthkournas@gmail.com (E.N.); ioannis.choreftakis@gmail.com (I.C.)
* Correspondence: antoniadis@dpem.tuc.gr; Tel.: +30-282-103-7293

Received: 2 July 2018; Accepted: 23 August 2018; Published: 3 September 2018

**Abstract:** Laser machining processes are a new entrant and a rapidly evolving type of non-conventional machining process which allows the machining of complex geometries with high precision, surface quality and productivity in a wide range of materials. Thus, the need for creating a method has emerged that will help the laser machine operator to select the optimal process parameters. In this study an experimental investigation of the effect of the process parameters on the effectiveness of the laser engraving process was held. The examined process parameters were namely the average output power, the repetition rate, and the scanning speed. For this purpose 126 experimental samples, with various combinations of process parameters using a nanosecond Nd:YAG DMG MORI Lasertec 40 laser machine on a SAE 304 stainless steel plate were made. The measured criteria which evaluated the effectiveness of the process were the removed material layer thickness and the material removal rate.

**Keywords:** laser engraving; laser machining; Nd:YAG laser

## 1. Introduction

The laser engraving process is a non-conventional, non-contact machining process which is rapidly evolving due to its ability to perform high-precision machining of complex geometries on a wide range of materials. The laser machining processes are gradually gaining ground over the conventional machining processes because of the advantage that there is no contact between a cutting tool and the workpiece. This means the elimination of problems like wear or failure of the cutting tools and the need to replace them periodically, which increases the production cost. Additionally, the laser machining processes facilitate the micromachining of complex geometries in workpieces with small dimensions, which would be impossible in many cases with conventional machining due to limitations from the accessibility and the minimum size of the cutting tool [1].

The basic working principle of the laser engraving process is that the laser beam machine generates laser pulses which provide a large amount of focused heat energy in the workpiece in order to cause the ablation of the material that has to be removed [2]. The manner and the amount of heat energy delivered to the material by the laser beam pulses can be changed by changing the basic process parameters, like the laser beam average power, the laser beam scanning speed, and the repetition rate of the laser pulses [3,4].

The main purpose of this project is the experimental investigation of how the laser beam machining process parameters (average output power, repetition rate, and scanning speed) affect the laser engraving of SAE 304 Stainless Steel by checking the resultant depth and the material removal rate. Laser machining processes are not as widespread as conventional ones because laser technology is at an early stage if we think that its evolution began in the early 1960s. Nevertheless, there are increasing number of research projects on laser machining providing solutions in tasks like the optimization of the generated surface quality, reducing energy consumption, increasing cutting speed, the machining of new materials, etc. Below are some research projects on laser beam machining.

Leone et al. [5] carried out an experimental research project to find the process parameters that play a decisive role in laser milling of aluminum oxide $Al_2O_3$ using a 30 W Q-switched Yb:YAG fiber laser machine, to explain the way that process parameters affect the interaction between laser beam and material, and to find the result of changing the process parameters on material removal rate and the surface quality. Kochergin et al. [6] carried out an experimental study whose main purpose was to examine the correlation between the process parameters of the laser machine and the produced surface micro geometry. For the experiments a 50 W Fiber Laser machine with a 50 µm laser spot diameter was used, and the tested materials were titanium BT1-0 and steel 12X18H10T. The process parameters tasted in the experiments were the laser pulse power, the frequency of the pulses and the pulses scanning speed. Angelastro et al. [7] conducted a study about investigating how the process parameters and building strategy affect the result of direct metal laser deposition of 227-F Colmonoy nickel alloy. They used a $CO_2$ laser machine with a 0.3 mm laser spot diameter and a 10 mm thick plate of AISI304 steel as substrate material. Their samples were characterized in terms of roughness, adhesion, micro and macrostructure, porosity, microhardness and relative density and they performed an Analysis of Variance (ANOVA) in order to investigate the influence of the process parameters on the quality of the parts. Casalino et al. [8] performed an experimental investigation and statistical optimization of the process parameters of the selective laser melting process of the 18Ni300 maraging steel by using a Nd:YAG laser machine. They examined the hardness, the mechanical strength, and the surface roughness of the samples and showed that those values are correlated positively to the part density. Manninen et al. [9] investigated how the laser pulse length affects laser engraving effectiveness of 304 stainless steel. They used a 20 W Ytterbium Fiber Laser for the experiments. The only process parameter tested was the pulse duration in a range between 4 and 200 ns. The effectiveness of the laser engraving processes was checked by the following factors: high material removal rate, good visual quality of the machined surface, and low material temperature during the process.

## 2. Materials and Methods

### 2.1. Laser Engraving Process

In the laser engraving process a focused laser beam is scanned over the workpiece. The energy of each laser pulse is absorbed by the workpiece and this heats the material, causing melting and finally vaporizing to a gas [10,11]. The phase transition from solid to vapor is referred to as an ablation process. As the evaporated material is ejected there is a material removal which finally gives the thickness of the single removed layer. By removing multiple layers of material with a different scanning pattern for each one, a 3D shaped surface structure can be produced [12].

Figure 1a presents a schematic illustration of the laser engraving process. The laser beam spot scans the workpiece material surface in a specific, predefined way (scanning strategy) and laser beam pulses are generated periodically, causing the ablation and the removal of the target material.

The material removal is affected by the characteristics of laser beam, the properties of the workpiece and the way they both interact [13,14]. The workpiece properties depend on the material and the geometry, with the most important being density, melting-vaporizing temperatures, specific heat capacity, heat conductivity, latent heat of melting-evaporation, and absorptivity-reflectivity in solid-melting states [15]. The laser beam can be characterized by the laser machine parameters, such as laser type, wavelength, laser spot diameter, pulse duration, and the process parameters, such as average output power P, repetition rate F and scanning speed V.

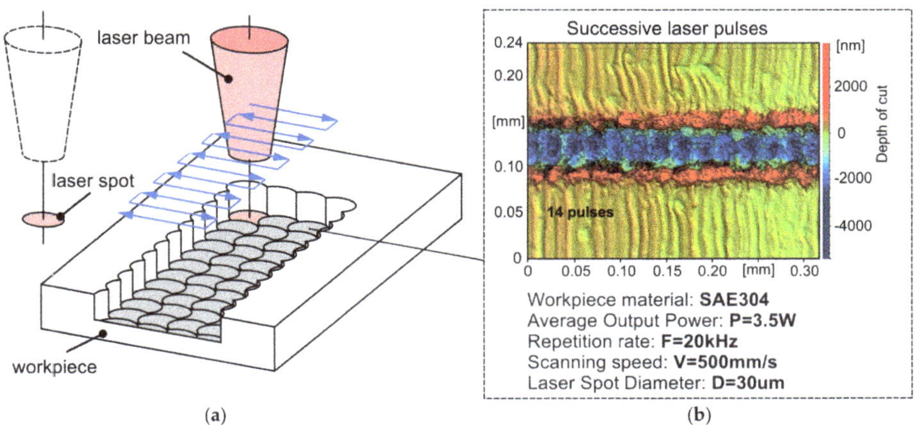



(a)  (b)

**Figure 1.** Laser engraving process (**a**) Schematic illustration of the laser engraving process; and (**b**) engraving depth of 14 successive laser beam pulses in a SAE304 stainless steel workpiece.

Figure 1b shows a picture of the removed material depth caused by 14 successive laser beam pulses in a SAE304 stainless steel workpiece. The picture was taken using a ContourGT-K 3D Optical Microscope. The engraving was performed with a DMG MORI Lasertec 40 machine with a nanosecond Nd:YAG lamp, 1064 nm wavelength and 30 μm laser spot diameter. The process parameters were average output power P = 3.5 W, repetition rate F = 20 kHz and scanning speed V = 500 mm/s.

## 2.2. Experimental Work

The main objective of this work is to investigate how the laser process parameters affect the removed material layer thickness and the material removal rate in the laser engraving process of AISI 304 stainless steel. For this purpose, experiments were carried out machining 4 mm × 4 mm square samples with a constant number of removed layers (50) with a total of 126 combinations of the three process parameters: average power P, repetition rate F, and scanning speed V. The experiments were performed using a DMG MORI Lasertec 40 machine and a 5 mm thickness SAE 304 stainless steel plate was used as the workpiece material. Figure 2a presents the exterior view of the machine while Figure 2b shows the table on which the workpiece is seated and the mirrors which direct the laser beams.

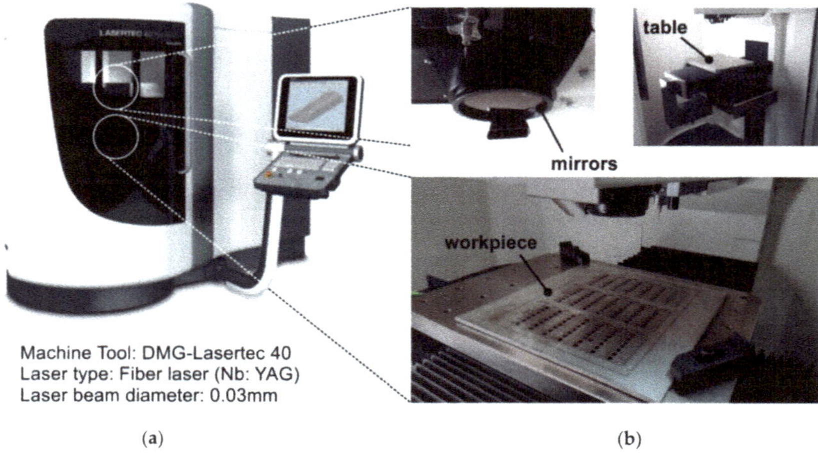



**Figure 2.** Laser engraving machine DMG MORI Lasertec 40 (**a**) Exterior view; and (**b**) interior view.

The process parameters were varied as follows: six levels for scanning speed V (200, 300, 400, 500, 600, 700 mm/s), seven levels for the repetition rate F (20, 30, 40, 50, 60, 70, 80 kHz), and three levels for the average power P (8, 12, 16 W). The scanning strategy used on each layer was cross-hatching, as shown in Figure 3. The track distance Td is the distance between two successive laser beam tracks, and the hatching distance Hd is the distance between two successive laser beam pulses in the same track. The track distance Td was set the same as the hatching distance Hd for the experiments and calculated by the following equation [16]:

$$\mathbf{Hd} = \frac{\mathbf{V}}{\mathbf{F}} \text{ , } \mathbf{Td} = \mathbf{Hd} \tag{1}$$

where: **Hd** is the hatching distance; **V** is the scanning speed; **F** is the repetition rate; **Td** is the track distance.

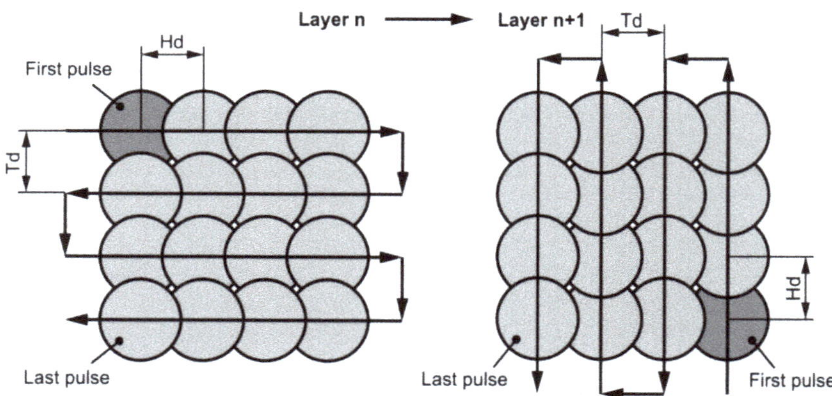



**Figure 3.** Laser beam scanning strategy.

The overlap degree Od between two following laser pulses is a useful value. It shows how two successive laser beam pulses overlap each other and can be calculated from the hatching distance Hd and the laser spot diameter D by the following equation [16]:

$$Od = \left(1 - \frac{Hd}{D}\right)\%$$ (2)

where Od is the overlap degree; Hd is the hatching distance; D is the spot diameter.

The removed material layer thickness Dz for each square sample was calculated by dividing the whole square depth (measured using the laser machine probing system) by the number of layers were performed (50 layers) as shown in the following equation [16]:

$$Dz = \frac{Dzn}{n}$$ (3)

where Dz is the removed material layer thickness; Dzn is the whole square depth; n is the total number of layers.

The material removal rate DV, the value that will determine the optimal processing parameters can be calculated for each combination of process parameters, be the following equation:

$$DV = V * Td * Dz$$ (4)

where DV is the material removal rate; V is the scanning speed; Td is the track distance; Dz is the removed material layer thickness.

## 3. Results

Figure 4 shows the produced workpieces with the different process parameters that were used.

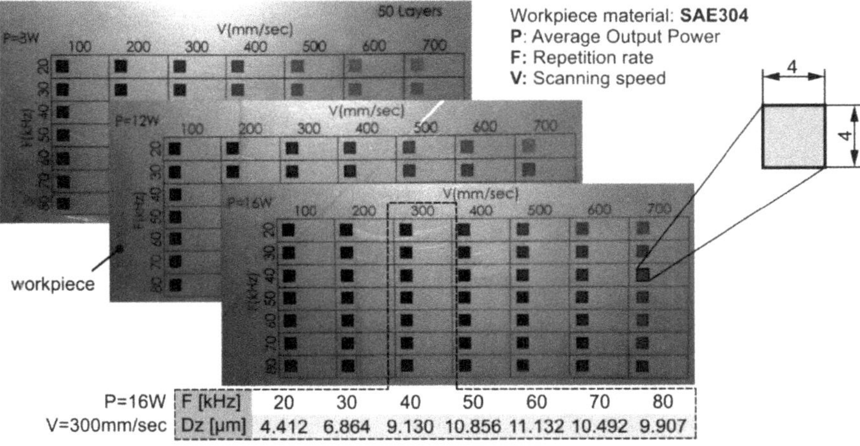


| P=16W | F [kHz] | 20 | 30 | 40 | 50 | 60 | 70 | 80 |
|---|---|---|---|---|---|---|---|---|
| V=300mm/sec | Dz [µm] | 4.412 | 6.864 | 9.130 | 10.856 | 11.132 | 10.492 | 9.907 |

**Figure 4.** Experimental samples.

### 3.1. Layer Thickness Dz

A laser machine probing system was used to measure the depth Dzn. Then the removed material layer thickness Dz was calculated for each sample using Equation (3). The results are shown in Figure 5.

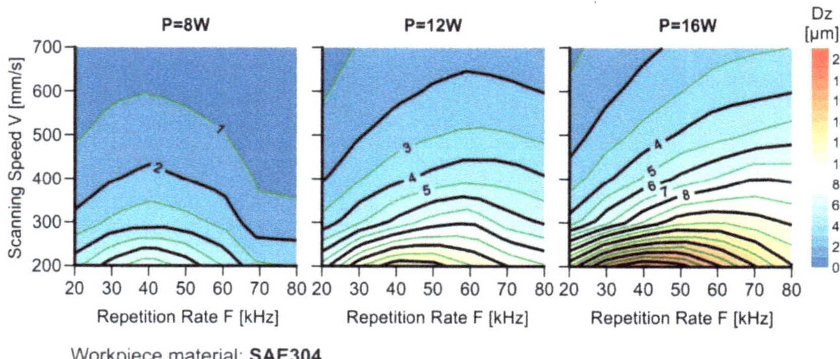


**Figure 5.** Removed material layer thickness Dz graphs.

Some conclusions can be drawn from the above graphs about the dependence of the removed material layer thickness Dz on the tested process parameters: average power P, repetition rate F, and scanning speed V. The maximum removed material layer thickness Dz is 19.4 µm and it is observed for process parameters: average power P = 16 W, repetition rate F = 40 kHz, and scanning speed V = 200 mm/s.

Independently of the applied average power P the maximum removed material layer thickness Dz is observed for repetition rate F = 40 kHz and scanning speed V = 200 mm/s, and it is 7.8 µm for P = 8 W, 12.9 µm for P = 12 W, and 19.4 µm for P = 16 W. As far as the average power P is concerned, it is observed that increasing the average power P while keeping the repetition rate F and the scanning speed V stable increases the removed material layer thickness Dz. As for the scanning speed V, it is observed that decreasing the scanning speed V while keeping the repetition rate F stable increases the removed material layer thickness Dz independently of the applied average power P. With regard to the repetition rate F, three cases are distinguished, depending on the applied average power P. For the low average power P = 8 W, the removed material layer thickness Dz maximizes at the repetition rate F = 40 kHz independently of the scanning speed V. For the medium average power P = 12 W, maximizing the removed material layer thickness Dz while keeping the scanning speed V stable requires a repetition rate F in the range between 40 kHz and 60 kHz, starting with the value of F = 40 kHz for low scanning speeds and reaching the value of F = 60 kHz for high scanning speeds. For the high average power P = 16 W the same behavior is observed as for the medium average power but the repetition rate F range at this time is between 40 kHz and 80 kHz, starting with the value of F = 40 kHz for low scanning speeds and reaching the value of F = 80 kHz for high scanning speeds.

*3.2. Material Removal Rate DV*

The material removal rate DV of each square can be calculated using Equation (4). The results are shown in Figure 6.

Some conclusions can be drawn from the above graphs about the effect of the tested process parameters on the material removal rate DV. The maximum material removal rate DV is 0.0261 mm$^3$/s and it is observed for process parameters: average power P = 16 W, repetition rate F = 20 kHz, and scanning speed V = 200 mm/s. Independently of the applied average power P, the maximum material removal rate DV is observed for repetition rate F = 20 kHz. For P = 8 W DV is 0.0118 mm$^3$/s at scanning speed V = 600 mm/s, for P = 12 W it is 0.0172 mm$^3$/s at scanning speed V = 700 mm/s, and for P = 16 W it is 0.0261 mm$^3$/s at scanning speed V = 200 mm/s. As far as the average power P is concerned it is observed that increasing the average power P while keeping the repetition rate F and the scanning speed V stable increases the material removal rate DV. As for the scanning speed V, two cases are distinguished, depending on the applied average power P. At the low and medium average powers P = 8 W and P = 12 W, it is observed that increasing the scanning speed V while keeping the repetition rate F stable increases the material removal rate DV. At the high average power P = 16 W, for a repetition rate F in the range between 20 kHz and 40 kHz the material removal rate DV increases by decreasing the scanning speed V, while for a repetition rate F in the range between 40 kHz and 80 kHz the material removal rate DV increases by increasing the scanning speed V. With regards to the repetition rate F it is observed that decreasing the repetition rate F while keeping the scanning speed V stable increases the material removal rate DV. However, a differentiation presents itself in the case of high average power P = 16 W for scanning speed V between 400 mm/s and 700 mm/s: by decreasing the repetition rate F at the beginning, the material removal rate DV increases until it reaches its maximum value for F = 40 kHz, and then a downward trend is observed.

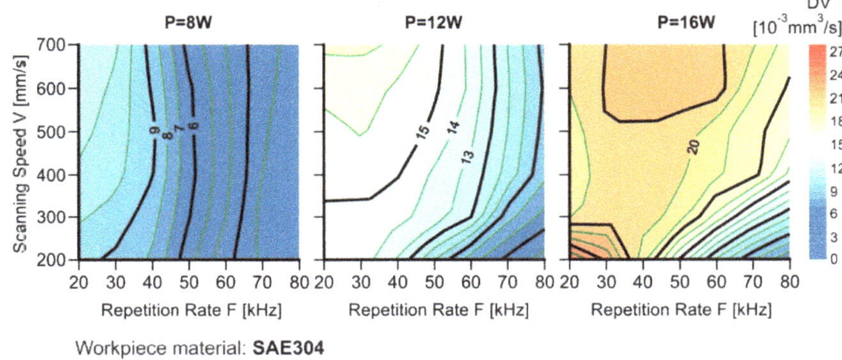


**Figure 6.** Material removal rate DV graphs.

## 4. Discussion

The purpose of this paper was to investigate the effect of the main laser machining process parameters (average output power, the repetition rate, and the scanning speed) on the effectiveness of the process. For this reason, a set of 126 experimental samples with various combinations of process parameters was made on a SAE304 stainless steel plate. The laser machine which was used was DMG MORI Lasertec 40. For each experimental sample, the removed material layer thickness was measured and the material removal rate was calculated; these were the criteria for the evaluation of the effectiveness of the process.

From the analysis of the experimental results many conclusions have been drawn which were analyzed in a previous section. In general, increasing the average output power P and decreasing the scanning speed V results in increasing the removed material layer thickness Dz. Moreover, increasing the average output power P, increasing the scanning speed V and decreasing the repetition rate F results in increasing the material removal rate DV.

**Author Contributions:** Conceptualization, E.N. and A.A.; Methodology, E.N.; Formal Analysis, E.N.; Investigation, I.C.; Writing-Original Draft Preparation, E.N.; Writing-Review & Editing, E.N.; Visualization, E.N.; Supervision, A.A.; Project Administration, A.A.

**Acknowledgments:** This research has been co-financed by the European Union and Greek national funds through the Operational Program Competitiveness, Entrepreneurship and Innovation, under the call RESEARCH-CREATE-INNOVATE (project code: T1EDK-01980)

## References

1.  Heyl, P.; Olschewski, T.; Wijnaendts, R.W. Manufacturing of 3D structures for micro-tools using laser ablation. *Microelectron. Eng.* **2001**, *57–58*, 775–780. [CrossRef]

2. Dubey, A.K.; Yadava, V. Laser beam machining—A review *Int. J. Mach. Tools Manuf.* **2008**, *48*, 609–628. [CrossRef]

3. Kasman, S. Impact of parameters on the process response: A Taguchi orthogonal analysis for laser engraving. *Measurement* **2013**, *46*, 2577–2584. [CrossRef]

4. Campanelli, S.; Casalino, G.; Ludovico, A.; Bonserio, C. An artificial neural network approach for the control of the laser milling process. *Int. J. Adv. Manuf. Technol.* **2013**, *66*, 1777–1784. [CrossRef]

5. Leone, C.; Genna, S.; Tagliaferri, B.; Palumbo, B.; Dix, M. Experimental investigation on laser milling of aluminium oxide using a 30 W Q-switched Yb:YAG fiber laser. *Opt. Laser Technol.* **2016**, *76*, 127–137. [CrossRef]

6. Kochergin, S.A.; Morgunov, Y.A.; Saushkin, B.P. Surface manufacturing under pulse fiber laser. *Procedia CIRP* **2016**, *42*, 470–474. [CrossRef]

7. Angelastro, A.; Campanelli, S.L.; Casalino, G. Statistical analysis and optimization of metal laser deposition of 227-F Colmony nickel alloy. *Opt. Laser Technol.* **2017**, *94*, 138–145. [CrossRef]

8. Casalino, G.; Campanelli, S.L.; Contuzzi, N.; Ludovico, A.D. Experimental investigation and statistical optimization of the selective laser melting process of maraging steel. *Opt. Laser Technol.* **2015**, *65*, 151–158. [CrossRef]

9. Manninen, M.; Hirvimaki, M.; Poutiainen, I.; Salminen, A. Effect of Pulse Length on Engraving Efficiency in Nanosecond Pulsed Laser Engraving of Stainless Steel. *Metall. Mater. Trans. B* **2015**, *46*, 2129–2136. [CrossRef]

10. Campanelli, S.L.; Ludovico, A.D.; Bonserio, C.; Cavalluzi, P.; Cinquepalmi, M. Experimental analysis of the laser milling process parameters. *J. Mater. Process. Technol.* **2007**, *191*, 220–223. [CrossRef]

11. Mladenovic, V.; Panjan, P.; Paskvale, S.; Caliskan, H.; Poljansek, N.; Cekada, M. Investigation of the laser engraving of AISI 304 stainless steel using a response-surface methodology. *J. Teh. Vjesn. Tech. Gaz.* **2016**, *23*, 265–271.

12. Romolia, L.; Tantussib, F.; Fuso, F. Laser milling of martensitic stainless steels using spiral trajectories. *Opt. Lasers Eng.* **2017**, *91*, 160–168. [CrossRef]

13. Teixidor, D.; Ferrer, I.; Ciurana, J.; Ozel, T. Optimization of process parameters for pulsed laser milling of micro-channels on AISI H13 tool steel. *Robot. Comput. Integr. Manuf.* **2013**, *29*, 209–218. [CrossRef]

14. Svantner, M.; Kucera, M.; Houdkova, S. Possibilities of stainless steel laser marking. In Proceedings of the 21st International Conference on Metallurgy and Materials, Brno, Czech Republic, 23–25 May 2012.

15. Sobotova, L.; Demec, P. Laser marking of metal materials. *Mod. Mach. Sci. J.* **2015**, 808. [CrossRef]

16. Campanelli, S.L.; Casalino, G.; Contuzzi, N. Multi-objective optimization of laser milling of 5754 aluminum alloy. *Opt. Laser Technol.* **2013**, *52*, 48–56. [CrossRef]

*Article*

# Experimental Analysis of the Effect of Vibration Phenomena on Workpiece Topomorphy Due to Cutter Runout in End-Milling Process [†]

Constantine David [1], Dimitrios Sagris [1,*], Evlampia Stergianni [2], Christos Tsiafis [2] and Ioannis Tsiafis [2]

[1] Laboratory of Manufacturing Technology and Production Systems (MT-Lab), Mechanical Engineering Department, Technological Education Institute of Central Macedonia, 62124 Serres, Greece; david@teicm.gr

[2] Laboratory of Machine Tools and Manufacturing Engineering, Mechanical Engineering Department, Aristotle University of Thessaloniki, 54124 Thessaloniki, Greece; estergia@meng.auth.gr (E.S.); tsiafis@gmail.com (C.T.); tsiafis@auth.gr (I.T.)

* Correspondence: dsagris@teicm.gr

[†] This paper is an extended version of our conference paper published in Stergianni, E.; Sagris, D.; Tsiafis, C.; David, C.; Tsiafis, I. Influence Analysis of Micro-Milling Vibrational Phenomena on Workpiece Topomorphy. In Proceedings of the 9th International Congress on Precision Machining (ICPM 2017), Athens, Greece, 6–9 September 2017.

Received: 4 April 2018; Accepted: 13 June 2018; Published: 1 July 2018

**Abstract:** Profile end-milling processes are very susceptible to vibrations caused by cutter runout especially when it comes to operations where the cutter diameter is ranging in few millimeters scale. At the same time, the cutting conditions that are chosen for the milling process have a complementary role on the excitation mechanisms that take place in the cutting area between the cutting tool and the workpiece. Consequently, the study of milling processes in the case that a cutter runout exists is of special interest. The subject of this paper is the experimental analysis of the effect of cutter runout on cutter vibration and, by extension, how this affects the chip removal and, thereby, the workpiece topomorphy. Based on cutting force measurements correlated with the workpiece topomorphy under various cutting process parameters, such as the cutting speed, feed rate, and the axial cutting depth, some useful results are extracted. Hence, the effect of vibration phenomena, caused by cutter runout, on the workpiece topomorphy in end milling can be evaluated.

**Keywords:** end-milling; cutter runout; cutting force; workpiece topomorphy

## 1. Introduction

During the cutting process, it is very important to know, as precisely as possible, the relationship between the vibration behavior of the machine tool-cutting tool-workpiece system and the cutting conditions (cutting speed, feed rate and axial depth of cut) in order to achieve high workpiece surface quality [1,2].

This is especially the case when it comes to the use of contemporary computer numerical control (CNC) machine tools in micromachining processes [3], for which it is crucial to define a group of optimal cutting conditions aiming at:

- ensuring dynamic stability [4,5] of the cutting process without side vibration phenomena [6] with an unfavorable impact both on the workpiece surface quality and the cutting tool life;
- achieving efficient machine tool performance;

- avoiding rapid development of cutting tool wear;
- reducing processing time and consequently production costs.

Finding an appropriate combination of process parameters has been such a challenging problem that a great deal of research has been devoted to the topic. Generally, it is difficult to define optimal cutting conditions to fit every cutting process. However, it is feasible to find out proper cutting parameters for a specific combination of machine tool, cutting tool and workpiece material.

The aim of the paper is to investigate experimentally the effect of the tool runout both on the cutting forces and the resulting workpiece surface. The paper follows a previous one [7], which was dealing with the effect of vibration phenomena on the workpiece topomorphy in general. The current research puts particular emphasis on the effect of vibrations caused by cutter runout on the workpiece surface quality. This is even more important when it comes to micromachining [3,8,9].

Tool clamping is usually accompanied by cutter runout. Although the cutter runout may be reduced by careful clamping performed by skilled machinists, a residual runout exists in the micro scale.

In the case of a residual cutter runout, unequal cutting forces are developed on the tool flutes during each cutter revolution. One of the flutes, where the eccentricity appears, undertakes the main cutting work and, therefore, is over-loaded, whereas the other flutes remain under-loaded. Because of that higher stresses are developed on the rake surface and the principal cutting edge of the particular flute. Such a situation leads to early flute wear, which deteriorates the whole cutting process. Moreover, the process is susceptible to becoming unstable [1].

Additionally, the effect of the cutter runout should be taken into consideration in order to secure the surface quality in profile end milling [10]. In this case, the enhanced cutting forces induce tool bending, thereby worsening the surface finish.

To investigate the effect of cutter runout on the cutting forces and the workpiece surface texture, dedicated experiments were conducted evaluating the cutting forces and the workpiece surface topomorphy. In particular, by means of experimental results, the cutting conditions which affect the cutting process outcome were studied. In this sense, the purpose of this experimental study was to find out the effect of the cutter runout on the cutting mechanism, on the tool vibration, and sequentially on the workpiece surface quality.

## 2. Experimental Procedure

To measure the cutting forces, specific milling experiments were conducted on the 5-axis CNC machining center DECKEL MAHO MH600C (DECKEL MAHO GmbH, Pfronten, Germany) shown in Figure 1. Before the measurement of the cutting forces, the workpiece was roughened in order to achieve the necessary flatness thus ensuring a constant depth of cut through the entire machined surface. For the roughing process, under the usage of coolant, an end-mill cutting tool of 40 mm diameter was selected.

The cutting tool used in the experiments was manufactured by the OSAWA Company (Osaka, Japan) with type G2CS4, and is shown in Figure 2. It is a 4-flute end-mill cutter of 2 mm diameter. The cutting tool material is micrograin carbide with multilayer physical vapor deposition (PVD) high performance coating of 3500 HV hardness. Due to its excellent abrasion resistance and the fact that it can withstand higher temperatures, this cutting tool is suitable for working with or without lubrication at high cutting speeds. All experiments of the research were performed under dry machining conditions.

The cutting tool was mounted on the ISO40 Tool holder by the aim of a ER32 collet. By a typical tool clamping procedure, the cutter was set up with a radial eccentricity leading to cutter runout of 7 μm. The resulted cutter runout was measured using a specific tool measurement device (DMG Violinear) with 1 μm resolution. The result of the eccentric tool mounting is schematically explained in Figure 3a.

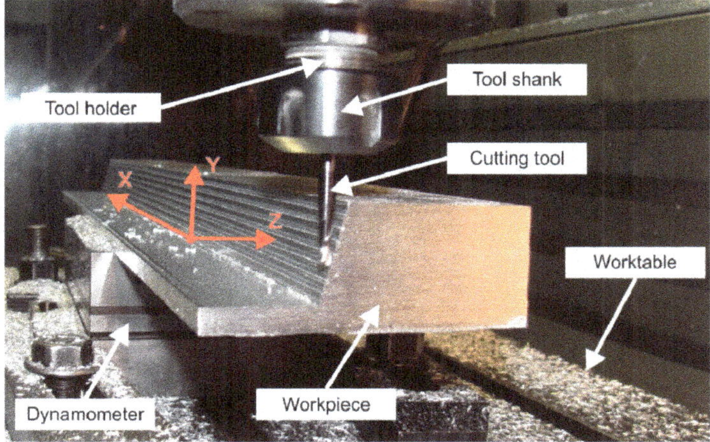



**Figure 1.** The experimental setup.

**Figure 2.** The cutting tool used in the experiments.

The cutting kinematics considering the cutter runout is depicted on the surface texture with a varied waviness profile (Figure 3c). Specifically, the height of the waviness profile is increased as well as the waviness length. The theoretical maximum roughness $R_t$ of the roughness profile is expressed by Equation (1) when the tool is ideally axisymmetric and, correspondingly, by Equation (2) in the case that a tool eccentricity exists [11,12]. From these equations it can be clearly seen that an increase in feed per tooth $f_z$ causes surface roughness growth, whereas an increase in tool diameter D, causes a decline in surface roughness.

$$R_{to} = \frac{D}{2} - \sqrt{\frac{D^2 - f_z^2}{4}} \simeq \frac{f_z^2}{4 \cdot D} \tag{1}$$

$$R_{te} = R - \frac{1}{2} \cdot \sqrt{4R^2 - f_z^2 - 2 \cdot e_r \cdot f_z - e_r^2} \simeq \frac{(f_z + e_r)^2}{4 \cdot D} \tag{2}$$

It is also worth noticing that the surface roughness parameters in the above equations often vary from the real surface roughness values, especially for low tool feed speed. One of the reasons of these discrepancies are cutter displacements related to the cutter's runout. Moreover, the tool's vibrations have an additional effect also.

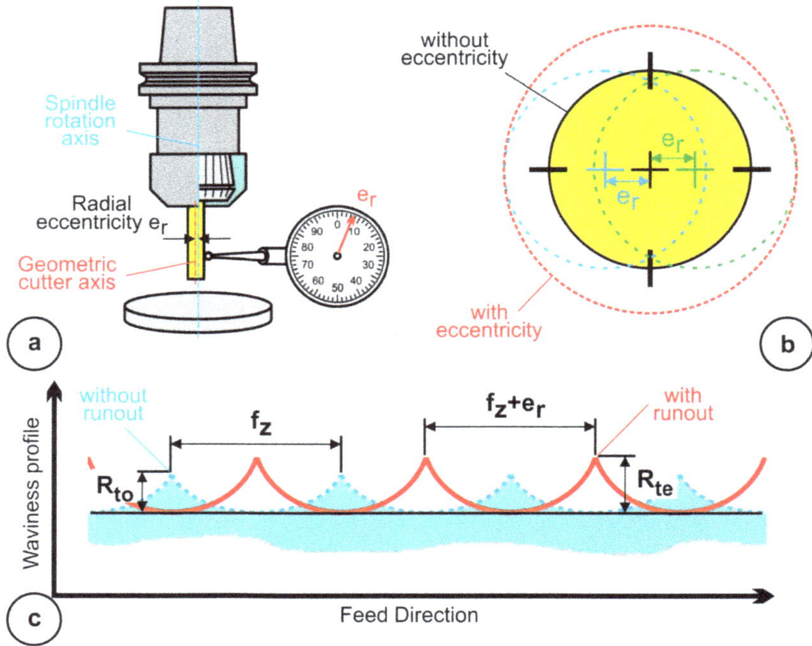



**Figure 3.** Representation of the workpiece waviness profile (**c**) due to tool radial eccentricity (**a**,**b**).

The parameter $e_r$ depicted in Figure 3b denotes the radial displacement between the cutting edges related to cutter runout. The cutter's static runout is mainly due to the offset between the position of the tool rotation axis and the spindle rotation axis. It is worth mentioning that in an end-milling process this radial cutter runout affects only the wall surface of the machined workpiece.

The material of the workpieces used for the experiments was aluminum alloy of class 7075 T651 (AlZn5.5MgCu). In Figure 4 the metallographic microstructure is shown. In the micrograph there are visible various phases granulation formed (Al-MgZn2, MgSi2, and phase contains Fe) in the Al matrix. This material has very high strength and, therefore, is used widely in applications in aerospace and in military industry.

In order to measure the cutting forces, metrological equipment was used, which consists of the stationary dynamometer type KISTLER 9257B (Kistler Group, Winterthur, Switzerland), the charge amplifier type KISTLER 5233A and an Analog/Digital converter of the National Instruments type NI PCI-MIO-16E (National Instruments, Austin, TX, USA) which was connected to a computer.

The three cutting force components Fx, Fy and Fz were measured in real time, whereby Fx component represents the feed force, Fy component represents the back force and Fz component represents the main cutting force (Figure 1). The measurements were taken with the aid of the inbuilt piezoelectric sensors. The output of the sensors is an electric charge (Q) linearly proportional to the force acting to the sensor. The charge amplifier intensifies this charge into a normalized voltage signal, which can then be digitized and acquired in the signal-processing system. The digital signals are subjected to a further processing in order to be evaluated both in time and frequency domain. The cutting parameters of the conducted experiments are shown in Table 1.

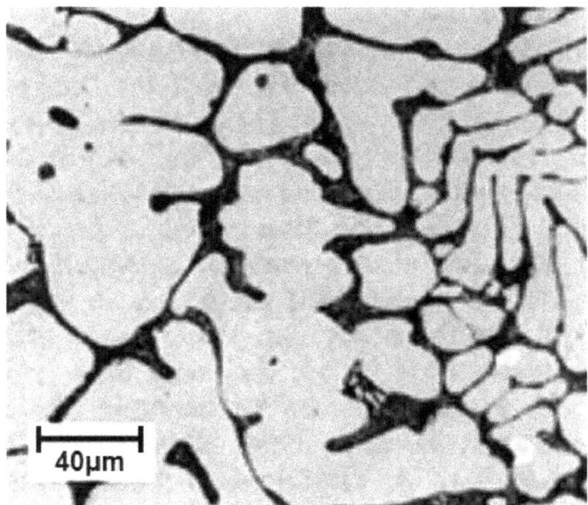


**Figure 4.** Metallographic structure of the aluminum alloy used in the experiments.

**Table 1.** Cutting conditions.

| Symbol | Parameter (Units) | | Value |
|---|---|---|---|
| S | Spindle Speed | (rpm) | 5500 |
| $V_c$ | Cutting Speed | (m/min) | 34.5 |
| f | Feed Speed | (mm/min) | 220/440 |
| $f_x$ | Feed per tooth | (mm/tooth) | 0.01/0.02 |
| $a_{xz}$ | Radial depth of cut | (mm) | 0.5 |
| $a_e$ | Axial depth of cut | (mm) | 0.2/0.4/0.6 |
| D | Cutter diameter | (mm) | 2 |

## 3. Experimental Results and Discussion

The cutting force components, after undergoing a low-pass filtering of 200 Hz, were recorded within a time frame of 0.04 s by a sampling rate of 25,000 samples/s. In Figures 5 and 6, typical measurement graphical interfaces are illustrated, showing the cutting parameters and the three cutting force components both in the time and frequency domains.

From all measurements, it is obvious that there is a direct correlation between the feed rate and the amplitude of the main cutting force (Fz) and the feed force (Fx). Regarding the passive force component (Fy), it is negligible.

Furthermore, taking into account that the cutting force is proportional to the axial cutting depth, it is confirmed that by working at higher material removal rates (MRR) more intense tool excitation is expected, leading to vibration phenomena capable of affecting the workpiece surface quality.

The effect of the tool runout on the cutting force and thereby on the vibration phenomena is clearly depicted in the Fast Fourier Transform (FFT) spectra of the cutting-force components. The tool rotation frequency amounts to 91.67 $s^{-1}$. From the cutting force spectra, we can see that only one cutter flute contributes to the cutting process at this frequency. This is more evident for the main cutting force (Fz) in relation to the feed force (Fx). Generally, the primer cutter flute located at the offset direction of the tool undertakes the main cutting work and the next flutes have less material to remove. This is demonstrated with the lower magnitude spectra peaks at $2\times$, $3\times$ and $4\times$ of the main tool-rotation frequency.

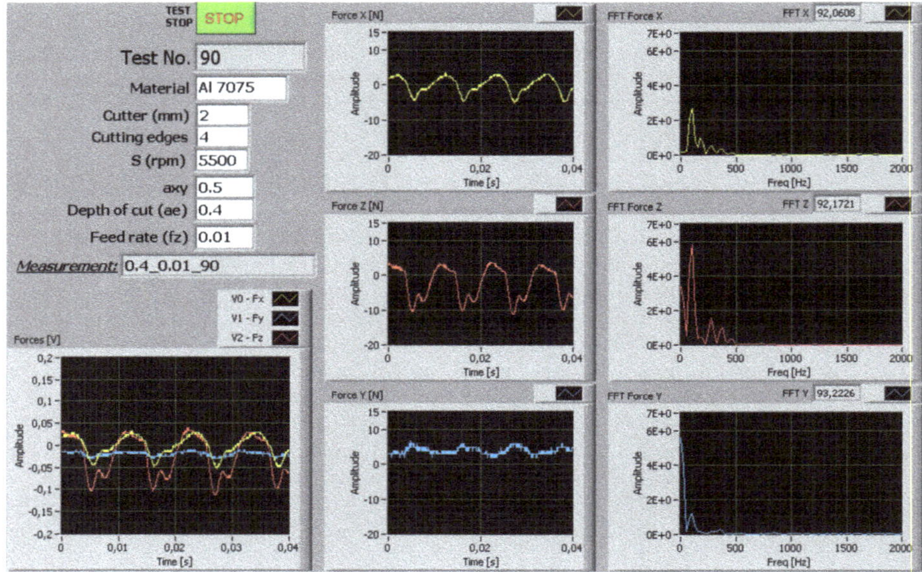

**Figure 5.** Cutting force components during milling process (feed rate 0.01 mm/tooth, axial depth of cut 0.4 mm).

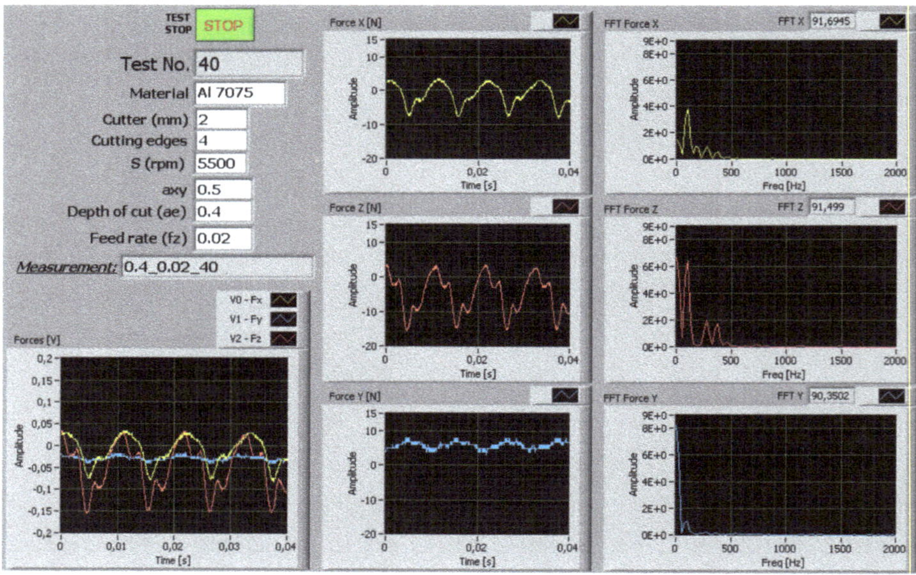

**Figure 6.** Cutting force components during milling process (feed rate 0.02 mm/tooth, axial depth of cut 0.4 mm).

In manufacturing, it is important to verify the topomorphy of the final workpiece in order to produce high-quality products. The influence of the cutting parameters in cutting processes and the workpiece surface is evaluated through the workpiece topomorphy and the roughness profile.

The combination of the cutting parameters has a direct impact on the cutting forces. This is due to the modification of the cutting process kinematics. As a result, the cutting process dynamics changes and, therefore, tool vibration excitations emerge.

In order to correlate the cutting parameters with the cutting force and thus the cutting dynamics the workpiece topomorphy was studied. For the quantitative and qualitative assessment of both workpiece topomorphy and the roughness profile, an optical profilometer of type VEECO NT1100 (Veeco, Tucson, AZ, USA) was used, which operates on the principle of white light interferometry. The measurement resolution in the depth direction of the workpiece topomorphy (perpendicular to the surface) amounted to less than 1 nanometer, while the resolution in the plane of the surface was about 0.4 microns. The resolution of the scanned areas by the white light interferometer was about 1 mm$^2$ (0.92 mm $\times$ 1.2 mm) for the $\times$10 magnification and 0.045 mm$^2$ (0.1853 mm $\times$ 0.2436 mm) for the $\times$50 magnification, captured by a 0.3 Mpixel sensor. The measurements were taken on the floor as well as on the wall of the machined surfaces. In Figure 7 the cutting tool paths in end milling are shown.

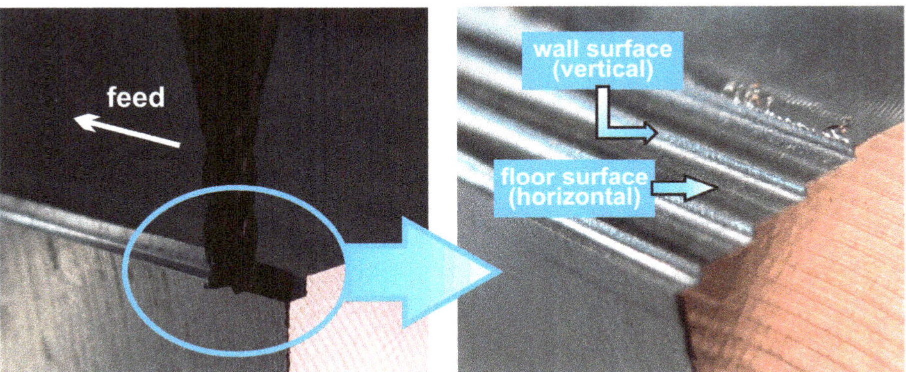


**Figure 7.** Cutting tool path and machined surfaces.

In Figure 8 the workpiece topomorphy and the roughness profile of the machined floor surface for feedrate 0.01 mm/tooth under various axial cutting depths are presented. The cutting tool traces are clearly depicted and according to the tool feedrate (0.04 mm/rev) the distance between two successive traces is approximately equal to 40 $\mu$m. In the digitized roughness profile, which refers to a length of 0.92 mm, a total of 23 tracks can be identified. This means that at every tool revolution just only one cutter flute path was imprinted on the surface texture. Apparently, each cutting flute takes part in the cutting process, but the eccentric one demonstrates higher force amplitude, which means that the specific flute cuts much more material. The traces of lower depth generated by the other flutes were removed because of the dominating flute path, leaving on the surface only one trace per tool revolution. Additionally, the tool deflection yielded an "axial runout", where the dominating flute was touching the workpiece more than the others, thus generating the observed bottom marks. Respectively, in Figure 9 the corresponding surface characteristics for feedrate 0.02 mm/tooth are shown. In this case, as expected, the distance between the successive tool traces amounted to 80 $\mu$m.

Table 2 shows the results of the maximum roughness value $R_t$ of the roughness profile concerning the machined floor surface under three axial cutting depths. Each of the results represents the mean value of $R_t$ ($\overline{R}_t$) for 10 profiles measured in the feed direction.

By increasing the axial cutting depth, the workpiece roughness becomes higher. At the same time an enhancement of the tool feedrate results in an additional increment of the workpiece roughness. This is expected since by increasing the volume of removed material the main cutting force increases too, leading to cutting tool deflection, which yields higher roughness.

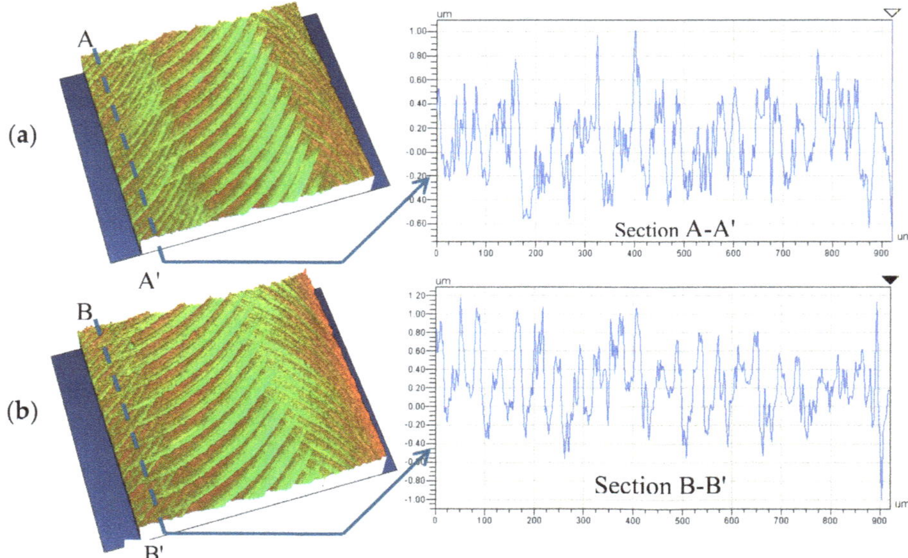

**Figure 8.** Topomorphy (zoom ×10) and roughness profile of the workpiece floor surface for 0.01 mm/tooth feedrate and 0.2 mm (**a**) and 0.4 mm (**b**) axial cutting depth.

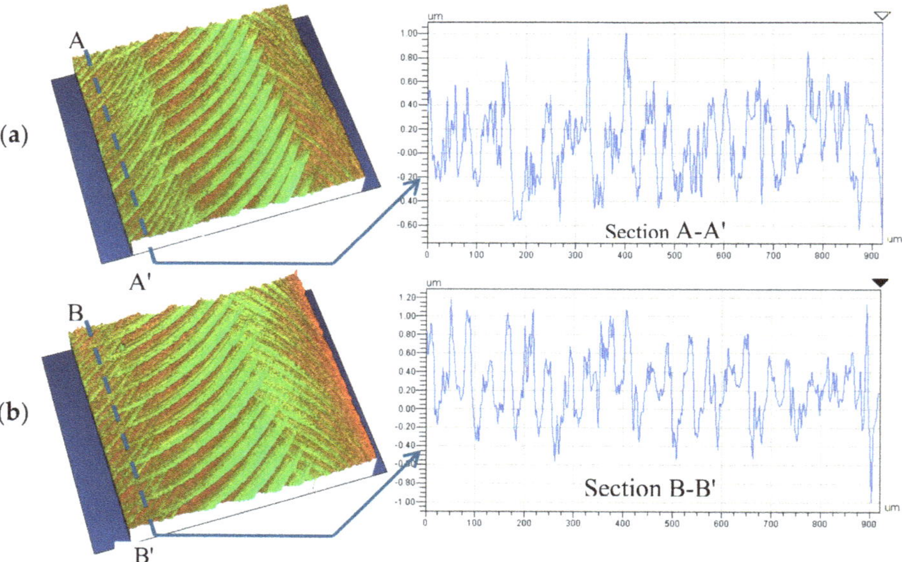

**Figure 9.** Topomorphy (zoom ×50) and roughness profile of the workpiece floor surface for 0.02 mm/tooth feedrate and 0.4 mm (**a**) and 0.6 mm (**b**) axial cutting depth.

**Table 2.** Maximum roughness profile value $R_t$ of the machined floor surface.

| Axial Cutting Depth (mm) | Feedrate 0.01 mm/Tooth | Feedrate 0.02 mm/Tooth |
|---|---|---|
| | Roughness Profile $\overline{R}t$ (µm) | |
| 0.2 | 1.68 | - |
| 0.4 | 2.18 | 2.78 |
| 0.6 | - | 3.13 |

In Figure 10 the topomorphy of the machined workpiece floor surface is presented in detail (zoom ×50). It can be observed that waviness is formed on the workpiece surface inside the cutting tool path (see figure detail). This waviness has occurred due to the tool vibration during the cutting process. The tool was subjected to high vibrations, as a result of the relatively high cutting forces, because of the tough cutting conditions of the specific experiment in relation to the others.

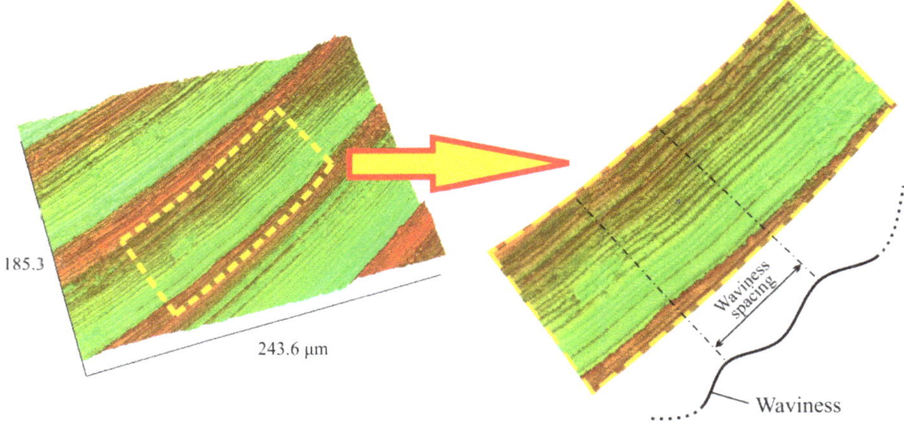


**Figure 10.** Workpiece topomorphy detail (zoom ×50) at the workpiece floor surface for tool feedrate of 0.02 mm/tooth and axial cutting depth of 0.6 mm.

In Figure 11, the workpiece topomorphy of the machined wall surface is illustrated for a tool feedrate of 0.02 mm/tooth and axial cutting depth of 0.2, 0.4 and 0.6 mm. The cutting tool traces due to the tool kinematics are shaped in the form of waves at each tool revolution. Observing the number of the waves at the examined length, the cutting tool feedrate can be validated. That means that in a length of about 0.9 mm, 11 waves appears with a wavelength of 80 µm, which is equal to the cutter feed per revolution (0.08 mm/rev). This means that just one cutting edge per tool revolution dominates in the cutting process because of the cutter runout effect. The waviness profile height depends on the axial depth of cut because of the superimposed cutter declination and cutter runout.

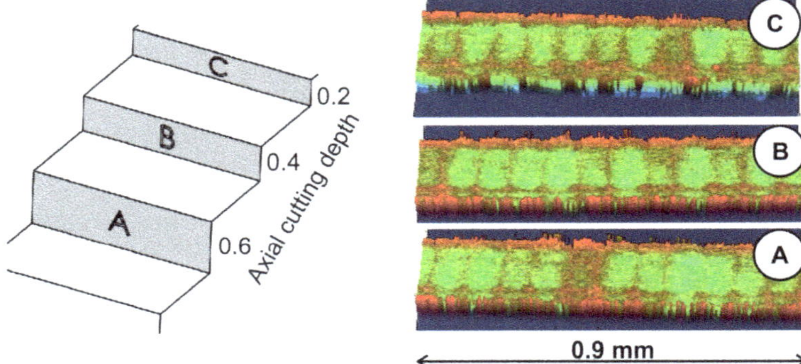


**Figure 11.** Workpiece topomorphy (zoom ×10) at the machined wall surface for tool feedrate of 0.02 mm/tooth.

## 4. Conclusions

In profile end-milling processes, the excitation mechanisms that take place in the cutting area between the cutting tool and the workpiece need to be carefully studied in order to ascertain the factors that affect the cutting operation. The present paper examines the effect of cutter runout on cutting forces and, thereby, on the workpiece topomorphy in cases where cutting tools of a few millimeters scale diameter are implemented in end-milling operations.

Specific experiments under various cutting conditions were conducted, whereby the cutting forces were recorded and appropriately processed. The results revealed that the cutting forces increased with the tool feedrate. Moreover, a rise of the axial cutting depth which resulted in a higher metal removal rate can cause more intense tool excitation, thus leading to vibration phenomena capable of affecting the workpiece surface finish.

In the case of the use of cutters with a residual runout, the topomorphy of the machined workpiece surface was also investigated. Increasing the axial cutting depth led to a higher surface roughness. Moreover, the surface roughness became worse with a higher tool feedrate. The experimental results have revealed that a direct correlation between the cutting conditions and both the workpiece topomorphy and the surface roughness exists.

In the case that a residual cutter runout is present, it has been verified that one cutting flute per tool revolution mostly takes part in the cutting process. Furthermore, evaluation of the machined workpiece topomorphy in end-milling process provides useful information about the effect of the vibration phenomena caused by cutter runout. To this end, the findings of this research can serve as useful suggestions for machinists who are confronting problems caused by tool runout.

**Author Contributions:** Conceptualization, C.D. and D.S.; Methodology, C.D. and D.S.; Software, D.S. and C.T.; Validation, C.D., D.S. and I.T.; Formal Analysis, C.D. and I.T.; Investigation, C.D., D.S. and E.S.; Resources, C.D., D.S., E.S., C.T. and I.T.; Data Curation, C.D. and D.S.; Writing-Original Draft Preparation, C.D., D.S. and E.S.; Writing-Review & Editing, C.D. and D.S.; Visualization, D.S.; Supervision, C.D.; Project Administration, C.D.

**Funding:** This research received no external funding.

**Conflicts of Interest:** The authors declare no conflict of interest.

## References

1.	Altintas, Y. *Manufacturing Automation—Metal Cutting Mechanics, Machine Tool Vibrations and CNC Design*, 2nd ed.; University of British Columbia: Vancouver, BC, Canada; Cambridge University Press: Cambridge, UK, 2012.

2.  David, C.; Sagris, D.; Stergianni, E.; Tsiafis, C.; Tsiafis, I. Experimental analysis of charter vibration on micro milling. *Ann. Constantin Brancusi Univ. Targu-Jiu Jurid. Sci. Ser.* **2016**, *4*, 210–216.
3.  Dornfeld, D.; Min, S.; Takeuchi, Y. Recent advances in mechanical micromachining. *CIRP Ann. Manuf. Technol.* **2006**, *55*, 745–768. [CrossRef]
4.  Quintana, G.; Ciurana, J. Chatter in machining processes: A review. *Int. J. Mach. Tools Manuf.* **2011**, *51*, 363–376. [CrossRef]
5.  Schmitz, T.L.; Smith, K.S. *Machining Dynamics Frequency Response to Improved Productivity*; Springer Science and Business Media, LLC: Berlin/Heidelberg, Germany, 2009.
6.  Altintas, Y. Research on metal cutting, machine tool vibrations and control. *J. Jpn. Soc. Precis. Eng.* **2011**, *77*, 470–471.
7.  Stergianni, E.; Sagris, D.; Tsiafis, C.; David, C.; Tsiafis, I. Influence Analysis of micro-milling vibrational phenomena on workpiece topomorphy. *Solid State Phenom.* **2017**, *261*, 77–84. [CrossRef]
8.  Rivière-Lorphèvre, E.; Filippi, E. Mechanistic cutting force model parameters evaluation in milling taking cutter radial runout into account. *Int. J. Adv. Manuf. Technol.* **2009**, *45*, 8. [CrossRef]
9.  Oliaei, S.N.B.; Karpat, Y. Experimental investigations on micro milling of stavax stainless steel. *Procedia CIRP* **2014**, *14*, 377–382. [CrossRef]
10. Matsumura, T.; Tamura, S. Cutting force model in milling with cutter runout. *Procedia CIRP* **2017**, *58*, 566–571. [CrossRef]
11. Koenig, W.; Klocke, F. *Fertigungsverfahren 1: Drehen, Fräsen, Bohren*; Springer-Verlag: Berlin/Heidelberg, Germany, 2002; ISBN 3-540-43304-X.
12. Wojciechowski, S.; Twardowski, P.; Wieczorowski, M. Surface texture analysis after ball end-milling with various surface inclination of hardened steel. *Metrol. Meas. Syst.* **2014**, *21*, 145–156. [CrossRef]

*Article*

# Prediction of Thrust Force and Cutting Torque in Drilling Based on the Response Surface Methodology

**Panagiotis Kyratsis** [1,*] , **Angelos P. Markopoulos** [2] , **Nikolaos Efkolidis** [1],
**Vasileios Maliagkas** [1] **and Konstantinos Kakoulis** [1]

[1]  Department of Mechanical Engineering & Industrial Design,
    Western Macedonia University of Applied Sciences, Kila Kozani GR 50100, Greece; nefko@teiwm.gr (N.E.);
    vmaliagkas@gmail.com (V.M.); kkakoulis@teiwm.gr (K.K.)
[2]  School of Mechanical Engineering, Section of Manufacturing Technology,
    National Technical University of Athens, Heroon Polytechniou 9, Athens 15780, Greece;
    amark@mail.ntua.gr
*  Correspondence: pkyratsis@teiwm.gr; Tel.: +30-2461-068-086

Received: 25 April 2018; Accepted: 31 May 2018; Published: 4 June 2018

**Abstract:** In this research, experimental studies were based on drilling with solid carbide tools and the material used was Al7075. The study primarily focused on investigating the effects of machining parameters (cutting speed, feed rate, diameter of the tool) on the thrust force ($F_z$) and the cutting torque ($M_z$) when drilling an Al7075 workpiece. The experimental data were analyzed using the response surface methodology (RSM) with an aim to identify the significant factors on the development of both the $F_z$ and $M_z$. The application of the mathematical models provided favorable results with an accuracy of 3.4% and 4.7%, for the $F_z$ and the $M_z$, respectively. Analysis of variance (ANOVA) was applied in order to examine the effectiveness of the model, and both mathematical models were established and validated. The equations derived proved to be very precise when a set of validation tests were executed. The importance of the factors' influence over the responses was also presented. Both the diameter of the cutting and the feed rate were found to be the factors of high significance, while cutting speed did not affect considerably the $F_z$ and $M_z$ in the experiments performed.

**Keywords:** Al7075; thrust force; cutting torque; response surface methodology

## 1. Introduction

A majority of products are directly or indirectly related to the metal cutting processes during their production processes. From the most commonly used processes, such as milling, turning, tapping and so forth, drilling constitutes the most commonly used. A significant number of holes are necessary for the assembly of different parts in order to be connected each other and develop the product itself.

Many researchers have developed empirical methodologies based on statistics, which can be used to learn the interactions between manufacturing factors. Response surface methodology (RSM) is one of the most widely used. RSM is a collection of an empirical modeling approach for defining the relationship between a set of manufacturing parameters and the responses. Different criteria are used and the significance of these manufacturing parameters on the coupled responses are examined. This methodology uses a number of techniques based on statistics, graphics, and mathematics, for both improving and optimizing a manufacturing process. Kumar and Singh [1] used ANOVA to investigate the material removal rate (MRR) and surface roughness of optical glass BK-7, which was drilled by a rotary ultrasonic machine. A number of manufacturing parameter combinations were used in order to effectively investigate the drilling process itself. The experimental process showed that the most influential factor was the feed rate, while the developed prediction models kept the error within 5% at a 95% confidence level. RSM was used by Balaji et al. [2] on the drilling process of Ti-6Al-4V alloy. Drilling parameters (i.e., spindle speed, helix angle, feed rate) were tested and the surface roughness, flank wear, and acceleration of drill vibration velocity (ADVV) were measured. A multi-response optimization was performed with a view to optimize the drilling parameters for minimizing the output in each case (surface roughness, flank wear, ADVV). Similarly, Balaji et al. [3] investigated the effects of helix angle, spindle speed, and feed rate on surface roughness, flank wear, and ADVV during the drilling process of AISI 304 steel with twist drill tools. The significant parameters were retrieved based on the RSM, while the optimum drilling parameters were identified based on a multi-response surface optimization method.

Furthermore, Nanda et al. [4] used RSM to measure the responses, such as material removal rate (MRR), surface roughness, and flaring diameter, with the three input parameters—pressure, nozzle tip distance, and abrasive grain size—on a borosilicate-glass workpiece with zircon abrasives. Boyacı et al. [5] developed a fuzzy mathematical model using a multi-response surface methodology based on the drilling process of PVC samples in an upright drill. Cutting parameters, such as cutting speed, feed rate, and material thickness, were tested for the minimization of surface roughness and cutting forces. RSM was also used by Ramesh and Gopinath [6] during drilling of sisal-glass fiber reinforced polymer composites to predict the influence of cutting parameters on thrust force. The adequacy of the models was checked using the analysis of variance (ANOVA). The results showed that the feed rate was the most influencing parameter on the thrust force, followed by the spindle and the drill diameter. Jenarthanan et al. [7] developed a mathematical model in order to predict how the input parameters (tool diameter, spindle speed, and feed rate) influence the output response (delamination) in machining of the ARALL composite using different cutting conditions. RSM was used for the analysis of the influences of entire individual input machining parameters on the delamination.

Rajamurugan et al. [8] developed empirical relationships to model thrust force in the drilling of GFRP composites by a multifaceted drill bit. The empirical relationships were developed by the response surface methodology, incorporating cutting parameters such as spindle speed (N), feed rate (f), drill diameter (d), and fiber orientation angle (q). The result was that the developed model can be effectively used to predict the thrust force in drilling of GFRP composites within the factors and their limits of the study. Rajkumar et al. [9] used RSM to predict the input factors influencing the delamination and thrust force on drilled surfaces of carbon-fiber reinforced polymer composite at different cutting conditions with the chosen range of 95% confidence intervals. Ankalagi et al. [10] used response surface methodology to analyze the machinability and hole quality characteristics in the drilling of SA182 steel with an HSS drill. They investigated the effects of machining parameters, such as cutting speed, feed rate, and point angle, on (a) thrust force, (b) specific cutting coefficient, (c) surface roughness, and (d) circularity error. The experimental methodology showed that thrust force, specific cutting coefficient, and surface roughness decrease with the increase in cutting speed, whereas the circularity error increases with increased cutting speed and feed, on micro-EDM drilling process parameters. Natarajan et al. [11] analyzed the effect of the manufacturing parameters (i.e., feed rate, capacitance, voltage) on machining a stainless-steel shim with a tungsten electrode. In all cases, the surface roughness, the MRR, and tool wear were measured. RSM helped the development of statistical models for multi-response optimization, based on the desirability function with an aim to determine the optimum manufacturing parameters. Finally, Kyratsis et al. [12] investigated the relationship of three manufacturing parameters (diameter of the tools, cutting speed, feed rate) on the cutting forces developed when drilling an Al7075 workpiece. The response surface methodology was the main tool for establishing the appropriate mathematical prediction models.

This paper presents a study of the Al7075 drilling process when using solid carbide tools with different diameters. A series of appropriately selected cutting speeds and feed rates were applied and different mathematical models for the prediction of the thrust force ($F_z$) and the cutting torque ($M_z$) were calculated. Analysis of variance (ANOVA) was used in order to validate the adequacy of the mathematical models and depict the significance of the manufacturing parameters. The novelty of this work is the development of mathematical models that can be used with high levels of confidence in order to predict the thrust force and torque expected during the drilling of Al7075, within a wide range of cutting parameters (cutting speed 50~150 m/min, feed rate 0.15~0.25 mm/rev). The optimum usage of cutting tools is something crucial for CNC users as it can affect the whole machining process.

## 2. Materials and Methods

### 2.1. Selection of Materials

Nowadays, the metal cutting process is a very important issue for the manufacturing sector. Al7075 is an aluminum alloy having zinc as the main alloying element. Due to its excellent properties (strength, density, thermal behavior), it is widely used in the manufacturing industry for vehicles. In addition, it provides the sector of aviation with a lighter, stronger aluminum component, as well as enabling safer, lighter, and more fuel-efficient vehicles. Furthermore, its ability to be highly polished makes it suitable for the mold tool manufacturing industry. The aforementioned reasons have led a number of researchers to focus their studies on these factors [13–15]. In this study, an Al7075 plate (150 mm × 150 mm × 15 mm) was selected for performing the experimental work. The material's mechanical properties and chemical composition is depicted in Table 1.

**Table 1.** Mechanical properties and chemical composition of Al7075.

| Mechanical Properties | | | | | |
|---|---|---|---|---|---|
| Young's Modulus | Density | Hardness, HV | Yield Strength | Tensile Strength | Thermal Conductivity |
| 72 GPa | 2800 kg/m$^3$ | 173 | 503 MPa | 572 MPa | 130 W/m-K |
| Chemical Composition | | | | | |
| Elements | Zn | Mg | Cu | Cr | Fe | Si | Mn | Ti | Al |
| Percentage | 6 | 3 | 2 | 0.3 | 0.6 | 0.5 | 0.4 | 0.3 | Balance |

### 2.2. Response Surface Methodology

RSM is a popular method for deriving mathematical models of manufacturing systems, based on the principles of statistics. The same method can be the basis for applying optimization as well. The input parameters are identified together with their value ranges in the application under study. Then, the experimental process is set. Finally, mathematical models are calculated for the responses measured in each case. The effects of all input parameters and their interactions on the response are analyzed and their importance is derived. Different statistical methodologies (i.e., analysis of means and variances) are developed in order to establish the accuracy and adequacy of the derived mathematical model.

### 2.3. Experimental Details

In this research, an Al7075 block was used as a stock material. A HAAS VF1 CNC machining center was used to perform all the drilling tests. The machining center is able to perform with continuous speed and feed control within its operational limits. During the tests, three cutting speeds ($V$) and three feed rates ($f$) were applied in combination with three cutting tools' diameters ($D$). Table 2 depicts the factors used in this research together with their levels.

**Table 2.** Machining factors and their levels.

| Factors | Notation | Levels | | |
|---|---|---|---|---|
| | | I | II | III |
| Cutting speed (m/min) | $V$ | 50 | 100 | 150 |
| Feed rate (mm/rev) | $f$ | 0.15 | 0.2 | 0.25 |
| Tool diameters (mm) | $D$ | 8 | 10 | 12 |

A mechanical vise was used for clamping the workpiece and a Weldon clamping device with high rigidity was used to clamp the drill. The type of cutting fluid was KOOLRite 2270 which was provided by the delivery system near to the cutting tool. A Kistler dynamometer type 9123 (capable of measuring four components at the same time) measured the thrust force ($F_z$) and torque ($M_z$) values in each case of a set of twenty-seven experiments, which were executed by using 27 different (one tool for each hole) non-through coolant solid carbide drill tools (Kennametal B041A, flute helix angle of 30 degrees). All the possible combinations of the manufacturing parameters were used (cutting speed, feed rate, tool diameter). Figure 1 depicts the steps of the research conducted. The main shape of the cutting tools and all the dimension details are illustrated in Figure 2.

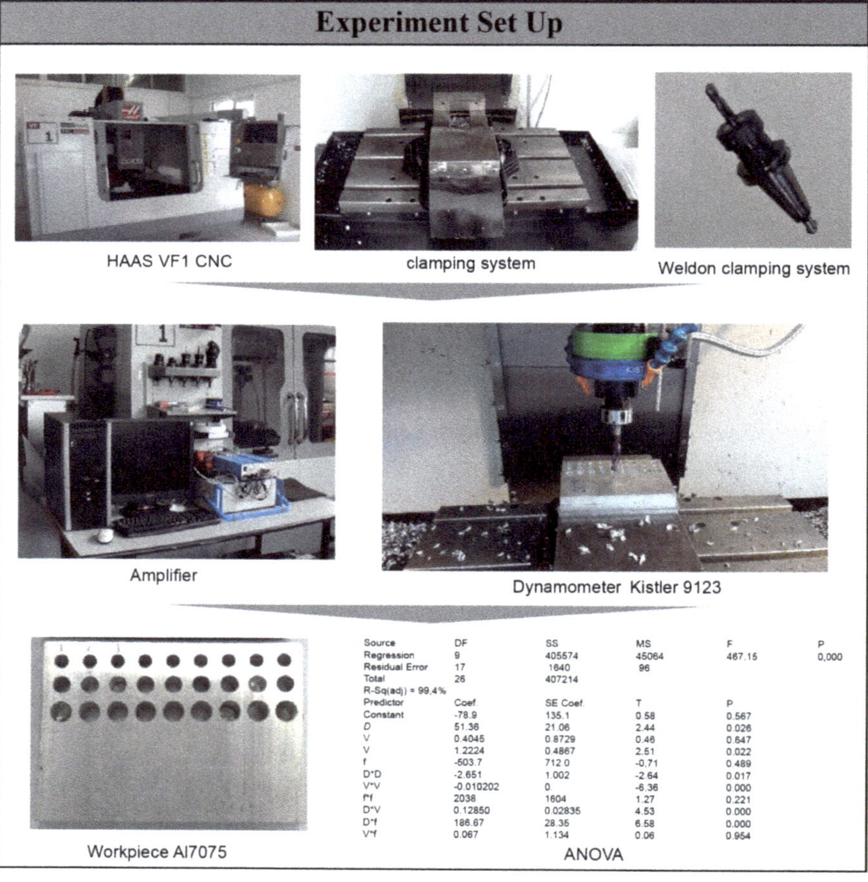


| Source | DF | SS | MS | F | P |
|---|---|---|---|---|---|
| Regression | 9 | 405574 | 45064 | 467.15 | 0.000 |
| Residual Error | 17 | 1640 | 96 | | |
| Total | 26 | 407214 | | | |
| R-Sq(adj) = 99.4% | | | | | |

| Predictor | Coef. | SE Coef. | T | P |
|---|---|---|---|---|
| Constant | -78.9 | 135.1 | 0.58 | 0.567 |
| D | 51.36 | 21.06 | 2.44 | 0.026 |
| V | 0.4045 | 0.8729 | 0.46 | 0.647 |
| V | 1.2224 | 0.4867 | 2.51 | 0.022 |
| f | -503.7 | 712.0 | -0.71 | 0.489 |
| D*D | -2.651 | 1.002 | -2.64 | 0.017 |
| V*V | -0.010202 | 0. | -6.36 | 0.000 |
| f*f | 2038 | 1604 | 1.27 | 0.221 |
| D*V | 0.12850 | 0.02835 | 4.53 | 0.000 |
| D*f | 186.67 | 28.35 | 6.58 | 0.000 |
| V*f | 0.067 | 1.134 | 0.06 | 0.954 |

**Figure 1.** The workflow used for the research.

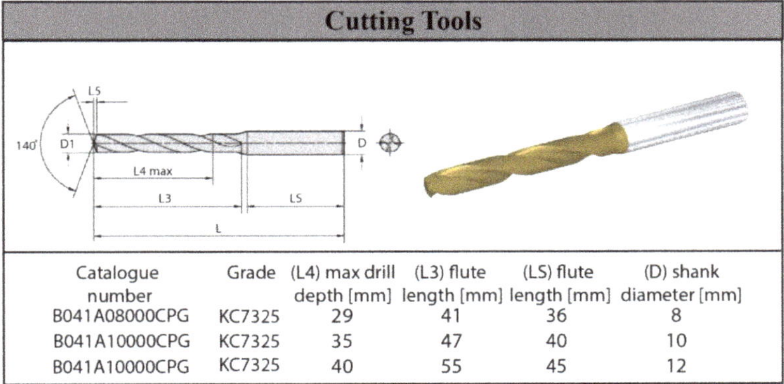



| Catalogue number | Grade | (L4) max drill depth [mm] | (L3) flute length [mm] | (LS) flute length [mm] | (D) shank diameter [mm] |
|---|---|---|---|---|---|
| B041A08000CPG | KC7325 | 29 | 41 | 36 | 8 |
| B041A10000CPG | KC7325 | 35 | 47 | 40 | 10 |
| B041A10000CPG | KC7325 | 40 | 55 | 45 | 12 |

**Figure 2.** Cutting tool geometry details.

Figure 3 demonstrates a sample graph of the measurements of thrust force and cutting torque. As mentioned, Kistler 9123 dynamometer has been used to measure the thrust force and cutting torque during the drilling process. Dynoware software (type 2825D-02) was used in order to monitor and record the values through a three-channel charge amplifier with a data acquisition system. During the tests, the thrust force and cutting torque were displayed graphically on the computer monitor and analyzed, enabling early error detection, and ensuring a steady state condition. As the sampling rate was high (approximately 10 kHz), the mean value was used as the final data in order to secure the value accuracy. The thrust force and cutting torque monitoring was developed as a means of enhancing the process monitoring capabilities.

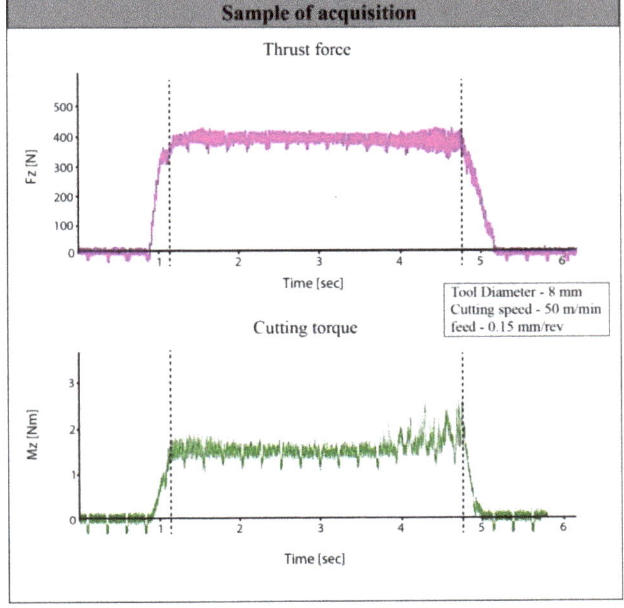



**Figure 3.** Experimental value (mean value) of thrust force and torque for a step drill with tool diameter of 8 mm, cutting speed of 50 m/min, and feed rate of 0.15 mm/rev.

Figure 4 illustrates all the experimental values derived from the dynamometer. It is obvious from the figure that when the diameter of the tool increases, both the $F_z$ and $M_z$ values increase. The same happens in the case of the feed rate. As the feed rate values increase, the $F_z$ and the $M_z$ increase, respectively. On the other hand, the different values of cutting speed do not noticeably affect their value in both cases.

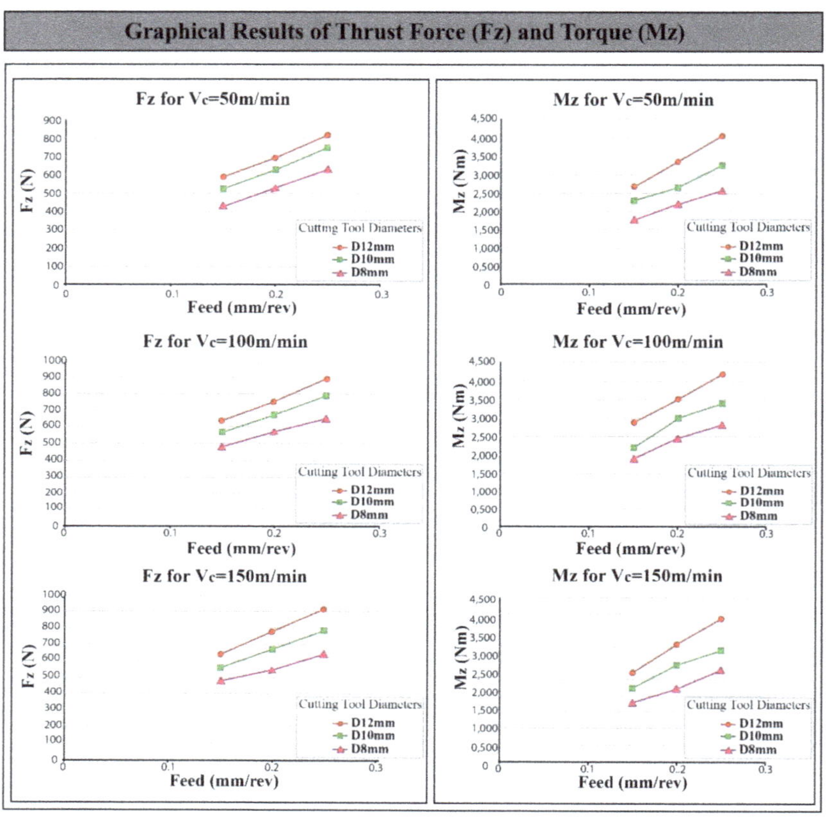


**Figure 4.** Experimental values derived from Kistler 9123.

## 3. Results

RSM was employed with a view to developing mathematical models for the $F_z$ and the $M_z$ in terms of cutting speed, feed rate, and cutting tool diameters. Least square fitting was used with the mathematical models in order to offer reliability. A full factorial strategy was applied and twenty-seven drilling experiments were performed, and both the $F_z$ and $M_z$ were modelled separately using polynomial mathematical models.

*RSM-Based Predictive Models*

The RSM is a precise tool used in order to examine the influence of a series of input variables on the response, when studying a complex phenomenon. The analysis of variance (ANOVA) was performed to check the adequacy and accuracy of the fitted models. MINITAB was used for the statistical analysis. The models produced used least square fitting and provided reliable mathematical modeling. The 2nd order nonlinear model with linear, quadratic, and interactive terms is indicated by the following equation:

$$Y = b_0 + b_1 X_1 + b_2 X_2 + b_3 X_3 + b_4 X_1^2 + b_5 X_2^2 + b_6 X_3^2 + b_7 X_1 X_2 + b_8 X_1 X_3 + b_9 X_2 X_3 \qquad (1)$$

where:

$Y$ is the response,

$X_i$ stands for the coded values, and

$b_i$ stands for the model regression coefficients.

According to the experimental values derived from Kistler 9123 (Figure 2) and the aforementioned mathematical model, the following equations were derived for the thrust force (in N) and the cutting torque (in Nm) respectively:

$$F_z = -79 + 51.4D + 1.22V - 504f - 2.65D^2 - 0.0102V^2 + 2038f^2 + 0.128DV + 187Df + 0.07Vf \, (\text{N})$$

and

$$M_z = 1.51 - 0.309D + 0.00236V + 1.06f + 0.016D^2 - 0.000037V^2 - 12.5f^2 + 0.000208DV + 1.33Df$$
$$+ 0.0213Vf \, (\text{Nm})$$

where:

$D$ is the diameter of the tool in mm,

$f$ is the feed rate in mm/rev, and

$V$ is the cutting speed in m/min.

The adequacy of the models is secured at a 0.05 level of significance. The validity of the developed models is proved by the use of ANOVA. The R-sq(adj) is very high in both cases (99.4% for the $F_z$ and 98.9% for the $M_z$). In addition, when a 0.05 level of significance is used, the main contributors of the models are those with a $p$-value less than 0.05. In the case of the $F_z$, these contributors are: $D$ ($p = 0.026$), $V$ ($p = 0.022$), $D^2$ ($p = 0.017$), $V^2$ ($p = 0.000$), $DxV$ ($p = 0.000$) and $Dxf$ ($p = 0.000$), while in the case of the $M_z$ the main contributors are: $D^2$ ($p = 0.048$), $V^2$ ($p = 0.007$), $Dxf$ ($p = 0.000$), $Vxf$ ($p = 0.023$).

The validity of the models is also proved from the values of $F$ and $p$. They indicate the significance of the mathematical model. The quality of the entire mathematical model can be proved by the $F$ value which considers all the manufacturing parameters at a time. The $p$ value depicts the probability of the manufacturing parameters having insignificant effect on the response. A large $F$ value signifies better fit of the mathematical model with the acquired data from the experiments. The derived values of F-ratio for the models of $F_z$ and $M_z$ (Tables 3 and 4) are calculated equal to 467.15 and 261.36, respectively. They are both higher than the standard tabulated values. A mathematical model is considered reliable when the $F$ value is high and the $p$-value is low (below 0.05 at a 0.05 level of significance).

Figures 5 and 6 both depict the main effect and interaction plots for the $F_z$ and $M_z$. In the interaction plot the interaction between diameter of the tool versus the cutting speed, and diameter of the tool versus the feed rate in all cases has been highlighted. As can be seen in Table 3, $V^2$ ($p = 0.000$), which implies that the square of the cutting speed affects the value of $F_z$ and $M_z$. This is more obvious when observing the main effect plots for $F_z$ and $M_z$ where the three means of the V depict a second order shape, compared to the rest of the mean effects plots that form almost straight lines. The same happens in the case of the cutting torque.

**Table 3.** ANOVA table for $F_z$ (thrust force).

| Source | Degree of Freedom | Sum of Squares | Mean Square | $f$-Value | $p$-Value |
|---|---|---|---|---|---|
| Regression | 9 | 405,574 | 45,064 | 467.15 | 0.000 |
| Residual Error | 17 | 1640 | 96 | | |
| Total | 26 | 407,214 | | | |
| R-Sq(adj) = 99.4% | | | | | |

| Predictor | Parameter Estimate Coefficient | Standard Error Coefficient | $t$-Value | $p$-Value |
|---|---|---|---|---|
| Constant | −78.9 | 135.1 | 0.58 | 0.567 |
| $D$ | 51.36 | 21.06 | 2.44 | 0.026 |
| $V$ | 1.2224 | 0.4867 | 2.51 | 0.022 |
| $f$ | −503.7 | 712.0 | −0.71 | 0.489 |
| $D{*}D$ | −2.651 | 1.002 | −2.64 | 0.017 |
| $V{*}V$ | −0.010202 | 0. | −6.36 | 0.000 |
| $f{*}f$ | 2038 | 1604 | 1.27 | 0.221 |
| $D{*}V$ | 0.12850 | 0.02835 | 4.53 | 0.000 |
| $D{*}f$ | 186.67 | 28.35 | 6.58 | 0.000 |
| $V{*}f$ | 0.067 | 1.134 | 0.06 | 0.954 |

**Table 4.** ANOVA table for $M_z$ (torque).

| Source | Degree of Freedom | Sum of Squares | Mean Square | $f$-Value | $p$-Value |
|---|---|---|---|---|---|
| Regression | 9 | 12.7870 | 1.4208 | 261.36 | 0.000 |
| Residual Error | 17 | 0.0924 | 0.0054 | | |
| Total | 26 | 12.8794 | | | |
| R-Sq(adj) = 98.9% | | | | | |

| Predictor | Parameter Estimate Coefficient | Standard Error Coefficient | $t$-Value | $p$-Value |
|---|---|---|---|---|
| Constant | 1.505 | 1.014 | 1.48 | 0.156 |
| $D$ | −0.3086 | 0.1581 | −1.95 | 0.068 |
| $V$ | 0.002357 | 0.003654 | 0.65 | 0.527 |
| $f$ | 1.057 | 5.345 | 0.20 | 0.846 |
| $D{*}D$ | 0.016014 | 0.007525 | 2.13 | 0.048 |
| $V{*}V$ | −0.00003691 | 0.00001204 | 3.07 | 0.007 |
| $f{*}f$ | −12.51 | 12.04 | −1.04 | 0.313 |
| $D{*}V$ | 0.0002075 | 0.0002128 | 0.97 | 0.343 |
| $D{*}f$ | 1.3317 | 0.2128 | 6.26 | 0.000 |
| $V{*}f$ | 0.021300 | 0.008514 | 2.50 | 0.023 |

Residual analysis was performed in order to test the models' accuracy in both cases. The result was that the residuals appear to be normally distributed. They follow the indicated straight lines (almost linear pattern) while at the same time they are distributed almost symmetrically around the zero residual value proving that the errors are normally distributed. For both cases, there is not considerable variation of the observed values around the fitted line. The discrepancy of a particular value from its predicted value is called the residual value. When the residuals around the regression line are small, it means that the mathematical model is of high accuracy. All the scatter diagrams of the $F_z$ and $M_z$ residual values versus the fitted values provide enough evidence that the residual values are randomly distributed on both sides of the provided graph. The same is true for the residual values versus the order that the experiments were conducted (Figure 7). There is no evidence of any particular pattern in the residual values and, thus, the models proved their efficiency.

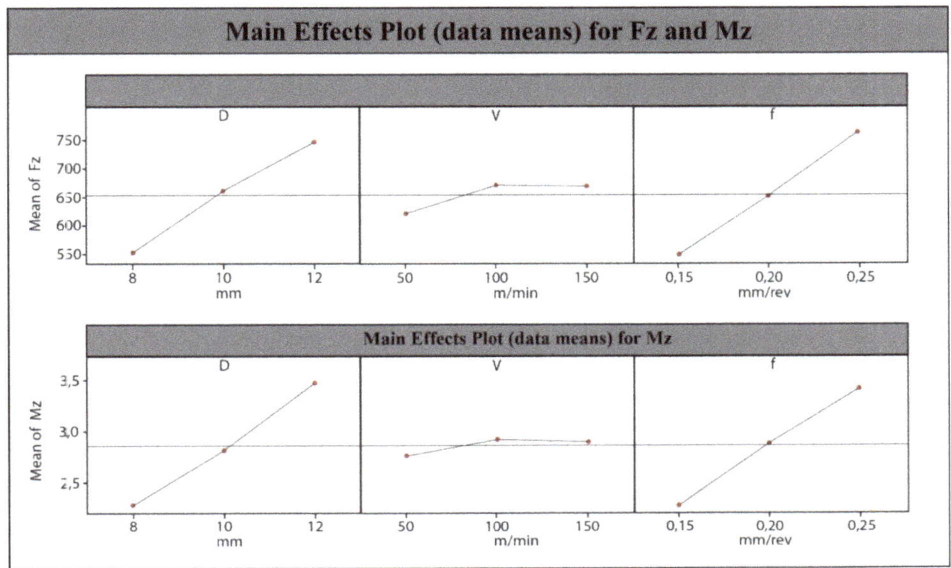

**Figure 5.** Main effects plot for $F_z$ and $M_z$.

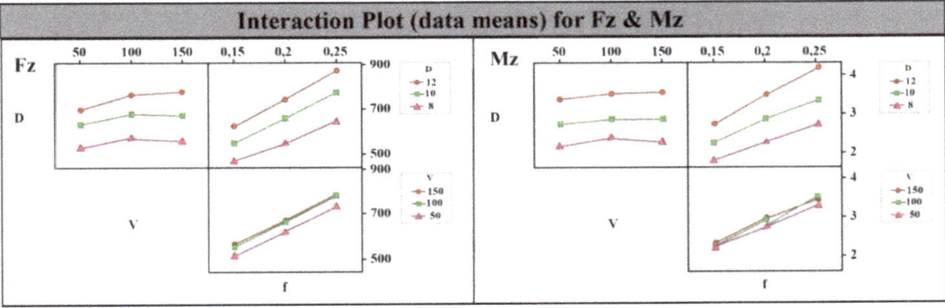

**Figure 6.** Interaction plot for $F_z$ and $M_z$.

That is the characteristic of well suited mathematical modeling and reliable data derived from the equations provided. The derived mathematical models can be considered as very accurate and can be used directly for predicting both the $F_z$ and the $M_z$ within the limits of the manufacturing parameters used (diameter of tool, feed rate, cutting speed). The accuracy achieved is very high when comparing the measured values and those calculated from the mathematical models (3.4% and 4.7%, respectively).

A set of experiments was conducted with a view to validate the given mathematical models for the prediction of the $F_z$ and $M_z$. Values of feed rate and cutting speed, within the range of experiments, were randomly selected ($V$: 70 m/min, $f$: 0.2 mm/rev) and tested with the three different cutting tools ($D$: 8-10-12 mm). The produced results can be considered satisfactory with less than 5.4% variation (Table 5). Especially, for the thrust force derived with the values ($D$: 10 mm, $V$: 70 m/min, $f$: 0.2 mm/rev), the variation was 0%.

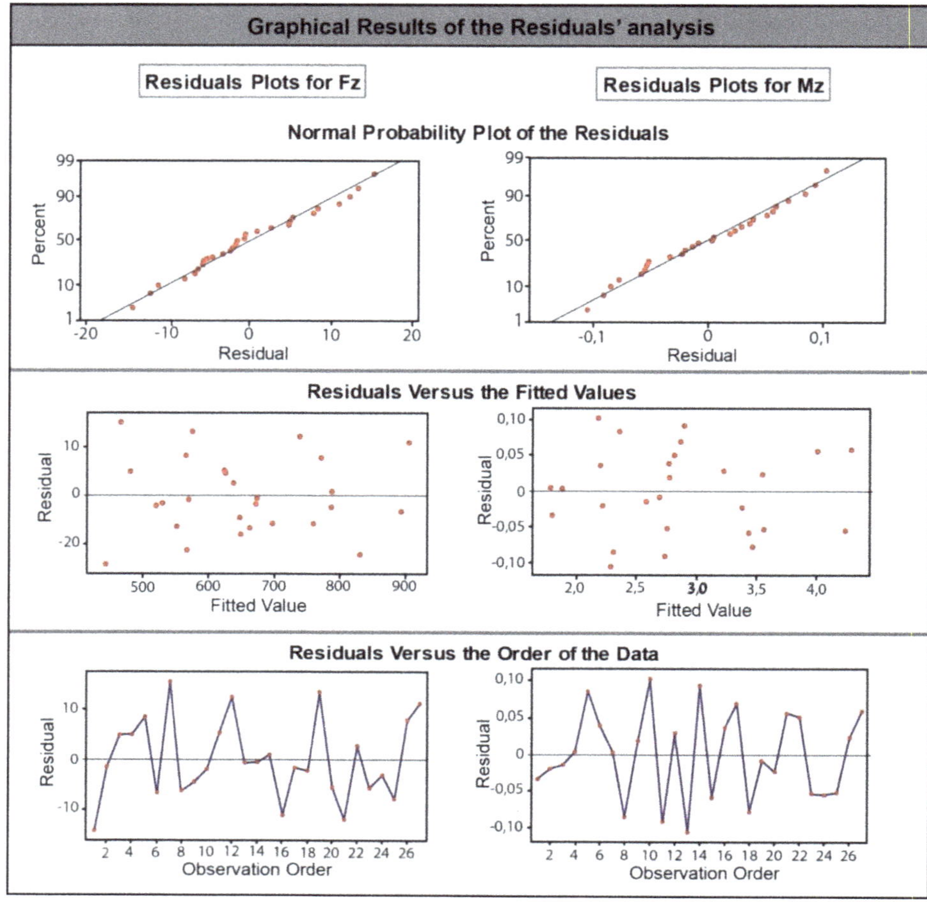



**Figure 7.** Residuals analyses for the $F_z$ and $M_z$.

**Table 5.** Experimental confirmation.

| Factors | | $Fz$ (N) | $Mz$ (Nm) |
|---|---|---|---|
| $D$: 8 mm, $V$: 70 m/min $f$: 0.2 mm/rev | Predicted | 707 | 1.899 |
| | Exp. Result | 692 | 1.855 |
| Variation % | | **2.1%** | **2.4%** |
| $D$: 10 mm, $V$: 70 m/min $f$: 0.2 mm/rev | Predicted | 875 | 2795 |
| | Exp. Result | 875 | 2955 |
| Variation % | | **0%** | **−5.4%** |
| $D$: 12 mm, $V$: 70 m/min $f$: 0.2 mm/rev | Predicted | 1060 | 4.129 |
| | Exp. Result | 1057 | 4.041 |
| Variation % | | **0.3%** | **2.2%** |

## 4. Conclusions

The aim of this study was the generation of mathematical models for the prediction of the thrust force ($F_z$) and cutting torque ($M_z$) related to the diameter of the drilling tool and other crucial manufacturing parameters (feed rate, cutting speed) during the drilling process. Research shows clearly that the prediction models sufficiently explain the relationship between the thrust force and cutting torque with the independent variables. A full factorial experimental process was executed under different conditions of the aforementioned parameters using an Al7075 workpiece and a full set of solid carbide tools. Response surface methodology was used as the base for the prediction of both the $F_z$ and the $M_z$ in a series of drilling operations with Al7075 as the material under investigation. The developed models were considered as very accurate for the prediction of the $F_z$ and $M_z$ within the range of the manufacturing parameters used. The outcomes proved that when using these models, the accuracy achieved was 3.4% and 4.7%, respectively. People working with CNC machines very often are faced with dilemmas about which cutting parameters are the most appropriate for the available cutting tools in order to treat materials, such as aluminum alloy 7075, which is suitable for a variety of specific applications in aerospace and chemical industries. Prediction of the thrust force and the cutting torque can lead to higher productivity when selecting manufacturing parameters due to the reduced wear effect on the drilling tool and the related energy consumption.

**Author Contributions:** Data curation, A.P.M.; formal analysis, N.E. and V.M.; investigation, P.K. and K.K.; methodology, P.K., A.P.M. and K.K.; validation, N.E. and V.M.

**Conflicts of Interest:** The authors declare no conflicts of interests

## References


1. Kumar, V.; Singh, H. Machining optimization in rotary ultrasonic drilling of BK-7 through response surface methodology using desirability approach. *J. Braz. Soc. Mech. Sci. Eng.* **2018**, *40*, 83. [CrossRef]
2. Balaji, M.; Venkata, K.; Mohan Rao, N.; Murthyd, B.S.N. Optimization of drilling parameters for drilling of TI-6Al-4V based on surface roughness, flank wear and drill vibration. *Measurement* **2018**, *114*, 332–339. [CrossRef]
3. Balaji, M.; Murthy, B.S.N.; Rao, N.M. Multi response optimization of cutting parameters in drilling of AISI 304 stainless steels using response surface methodology. *Proc. Inst. Mech. Eng. Part B J Eng. Manuf.* **2018**, *232*, 151–161. [CrossRef]
4. Nanda, B.K.; Mishra, A.; Dhupal, D.; Swain, S. Experimentation and optimization of process parameters of abrasive jet drilling by surface response method with desirability based PSO. *Mater. Today Proc.* **2017**, *4*, 7426–7437. [CrossRef]
5. Bcyacı, A.I.; Hatipoglu, T.; Balci, E. Drilling process optimization by using fuzzy-based multi-response surface methodology. *Adv. Ind. Eng. Manag.* **2017**, *12*, 163–172. [CrossRef]
6. Ramesh, M.; Gopinath, A. Measurement and analysis of thrust force in drilling sisal-glass fiber reinforced polymer composites. In *IOP Conference Series: Materials Science and Engineering*; IOP Publishing: Bristol, UK, 2017; Volume 197.
7. Jenarthanan, M.P.; Karthikeyan, M.; Naresh, N. Mathematical modeling of delamination factor on drilling of ARALL composites through RSM. *Multidscip. Model. Mater. Struct.* **2017**, *13*, 578–589. [CrossRef]
8. Rajamurugan, T.V.; Shanmugam, K.; Palanikumar, K. Mathematical Model for Predicting Thrust Force in Drilling of GFRP Composites by Multifaceted Drill. *Indian J. Sci. Technol.* **2013**, *6*, 5316–5324. [CrossRef]
9. Rajkumar, D.; Ranjithkumar, P.; Jenarthanan, M.P.; Sathiya Narayanan, C. Experimental investigation and analysis of factors influencing delamination and thrust force during drilling of carbon-fibre reinforced polymer composites. *Pigment Resin Technol.* **2017**, *46*, 507–524. [CrossRef]
10. Ankalagi, S.; Gaitonde, V.N.; Petkar, P. Experimental Studies on Hole Quality in Drilling of SA182 Steel. *Mater. Today Proc.* **2017**, *4*, 11201–11209. [CrossRef]
11. Natarajan, U.; Suganthi, X.H.; Periyanan, P.R. Modeling and Multiresponse Optimization of Quality Characteristics for the Micro-EDM Drilling Process. *Trans. Indian Inst. Met.* **2016**, *69*, 1675. [CrossRef]

12. Kyratsis, P.; Garcia-Hernandez, C.; Vakondios, D.; Antoniadis, A. Thrust force and torque mathematical models in drilling of Al7075 using the responce surface methodology. In *Book Design of Experiments in Production Engineering*; Davim, J.P., Ed.; Springer International Publishing: Berlin, Germany, 2016; pp. 151–164, ISBN 978-3-319-23837-1.
13. Pereira, R.B.D.; Leite, R.R.; Alvim, A.C.; Alvim, A.C.; Paiva, A.P.; Balestrassi, P.P.; Ferreira, J.R.; Davim, J.P. Multivariate robust modeling and optimization of cutting forces of the helical milling process of the aluminum alloy Al 7075. *Int. J. Adv. Manuf. Technol.* **2018**, *95*, 2691. [CrossRef]
14. Zou, X.L.; Yan, H.; Chen, X.H. Evolution of second phases and mechanical properties of 7075 Al alloy processed by solution heat treatment. *Trans. Nonferrous Met. Soc. Chin.* **2017**, *27*, 2146–2155. [CrossRef]
15. John, P.; Davis, R. Performance Study of Electrical Discharge Machining Process in Burn Removal of Drilled Holes in Al 7075. *Cogent Eng.* **2016**, *3*, 1–7. [CrossRef]

 *machines*

*Article*

# Influence of Hub Parameters on Joining Forces and Torque Transmission Output of Plastically-Joined Shaft-Hub-Connections with a Knurled Contact Surface

Lukáš Suchý [1,*], Erhard Leidich [1], Thoralf Gerstmann [2] and Birgit Awiszus [2]

[1] Professorship Engineering Design, Institute of Design Engineering and Drive Technology, Technische Universität Chemnitz, 09126 Chemnitz, Germany; erhard.leidich@mb.tu-chemnitz.de

[2] Professorship Virtual Production Engineering, Institute for Machine Tools and Production Processes, Technische Universität Chemnitz, 09126 Chemnitz, Germany; thoralf.gerstmann@mb.tu-chemnitz.de (T.G.); birgit.awiszus@mb.tu-chemnitz.de (B.A.)

* Correspondence: lukas.suchy@mb.tu-chemnitz.de; Tel.: +49-371-531-32446

Received: 27 February 2018; Accepted: 4 April 2018; Published: 9 April 2018

**Abstract:** A knurled interference fit is a machine part connection made by a plastic joining, which includes the advantages of commonly-used shaft-hub-connections. The combination of the friction and form fit, which are responsible for torque transmission, results in a higher power density than conventional connections. In this paper, parameter gaps are bridged with the aim of enhance the design calculation of the knurled interference fit. Experimental investigations on the shaft chamfer angle (100Cr6) and hub-diameter-ratio (AlSi1MgMn) were performed. The analytical approaches are developed for calculating the joining force and maximal torque capacity by accounting for experimentally investigated loss of load transmission at high hub-diameter-ratios and high shaft chamfer angles. The presented calculation approach is an accurate tool for the assessment of early machine designs of the knurled interference fit and helps to save from having to perform time-extensive tests.

**Keywords:** knurled interference fit; shaft-hub-connection; joining by plastic forming; torque capacity; joining force; drive train

## 1. Introduction

Turning drives are commonly used and important actuators in machines. Several principles for transmitting rotatory drive torque through machine parts from the source of power to the output side are established. Well-known principles are the key fit joint or the splined shaft fit, which belong to the group of so-called "positive fits" or "form fits". By contrast, the interference fit ("non-positive fit" or "friction fit"), where an oversized cylindrical shaft is pressed into the undersized cylindrical hub, transmits torque by joint pressure and is used in numerous applications, such as train axles. Combining these two well-known joining methods enables assembly of a splined shaft and hub with a cylindrical hole by axial plastic forming, bringing together the advantages of both principles and increasing the maximal transmissible torque (Figure 1). This so-called knurled interference fit (KIF) is fabricated by axial pressing the knurled shaft into a slightly undersized hole in the softer hub. The plastic forming of the counter profile in the hub eliminates tolerance errors and guarantees a homogenous force distribution over all tappets during operation. Furthermore, axial loads can be transmitted by the groove pressure in the connection.

The joining process of well-known interference push-in-connection can be estimated both according to the German standard for calculating interference fit DIN 7190 [1] as well as numerical methods from several investigations (e.g., [2,3]). In contrast to this investigation dealing global elastic-plastic relations with low deformation degree, a KIF connection is formed mainly locally at the contact groove with very high local deformation. Therefore the joining process of KIF cannot be reproduced with the presented methods.

In contrast to the conventional interference fit, the groove pressure decreasing separation frequency [4] in high-speed applications does not lead to transmission break down due to the form fit of the tappets. The knurl profile is understood according to German standard DIN 82 [5].

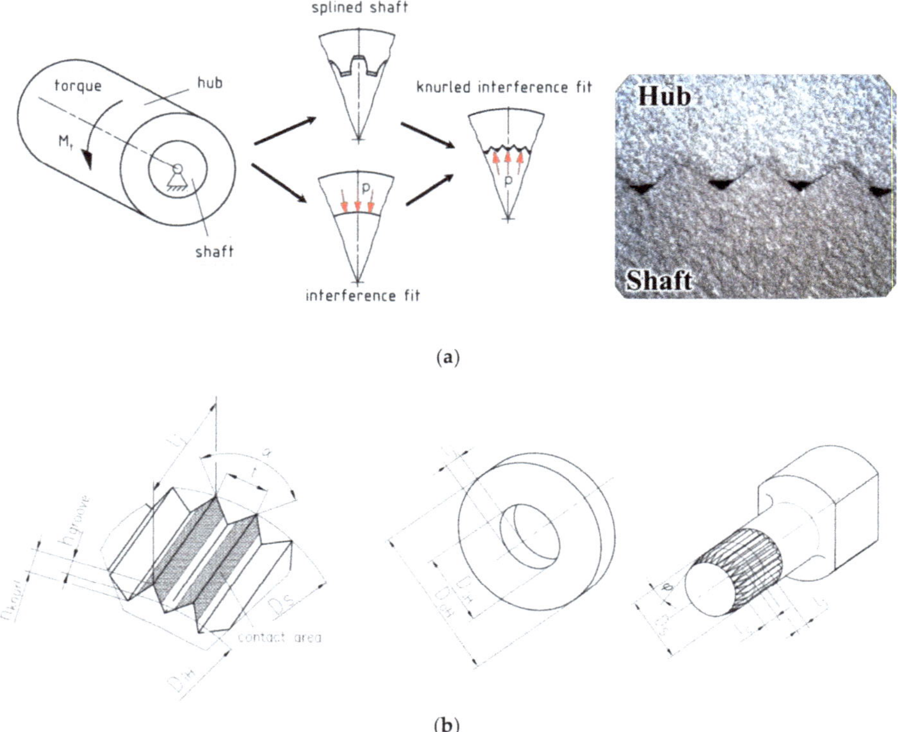


**Figure 1.** (**a**) Fusing of principles for torque transmission; and (**b**) geometry of knurled interference fit according to [5].

Equations describing this joint were recently published. Extensive recent overviews of past research can be found in [6–8], where the joining process and torsional load capacity are investigated. Nonetheless, the state of the knowledge is not sufficient for design of a universal KIF for a broad range of applications. Additionally, the valid patents in Europe lack detailed instructions for design and scientific explanations.

Lätzer [6] recently described the influence of different geometric parameters on joining and operating behavior, including important consequences of joining by cutting and forming. Fits joined by forming with shaft chamfer angles (SCA) $\phi < 60°$ are able to transmit 40% more torque than fits joined by cutting, where $\phi = 90°$. Additionally, an analytical approach for designing KIF based on operational state is introduced. Equations (1) and (2) shows Lätzer's approach [6] for estimation of necessary joining force and maximal transmittable torque:

$$T_{\tau_s} = D_S/2 \cdot t \cdot l_j \cdot i \cdot \tau_S\left(\varepsilon^{pl}_{KIF}\right) \tag{1}$$

$$F_J = A_{contact} \cdot \mu \cdot p\left(\varepsilon^{pl}_{KIF}\right) \cdot K_{QH} \tag{2}$$

With the exception of the hub-diameter-ratio (HDR), this formulation includes the geometry parameters of KIF (Figure 1). The terms $\tau_S\left(\varepsilon^{pl}_{KIF}\right)$ and $p\left(\varepsilon^{pl}_{KIF}\right)$, the equivalent plastic strain and the strain hardening, can be calculated by the Ludwik approach. Additionally, differentiation between the cutting and forming joining methods is performed. The results of this study are valid only for thick walled hubs of ductile aluminum material. The value of the hub-diameter ratio is the thickness rate of the hub and can be calculated for steady hub outer diameters using Equation (3):

$$Q_H = \frac{D_{iH}}{D_{oH}} \tag{3}$$

The interference $I_{geo}$ is the difference between the shaft diameter $D_S$ and the hub diameter $D_{iH}$:

$$I_{geo} = D_S - D_{iH} \tag{4}$$

Bader investigated the self-cutting knurling shaft-hub-connection [9] and determined the necessary hardness-ratio of the hub and shaft, which is calculated from different combinations of steel, brass, and aluminum. The minimal hardness ratio was 1.8. Furthermore, the first applicable empirical approaches were derived, but they lacked the universal validity for forming KIF.

Against this background, the central question that motivates the present research is the influence of other parameters on the maximal transmittable torque of the knurled interference fit, such as the hub thickness in combination with the shaft chamfer angle.

Further studies investigate forming KIF connections with different, and not standardized, geometries and tappet orientations [8]. Similar increasing load capacity tendencies were recorded in dependence of growing interference. Due to high influence of the different tappet geometry on the plastic forming during joining process, the quantitative comparability is not possible. Additionally, the shape development of the tappets is not introduced.

The lack of complete investigations on the influence of hub outside-diameter on joining and torsional load force make this research necessary. The loss of stiffness of thin walled hubs will change the maximal possible load capacity of the KIF. The validity condition of this statement is the failure of the knurls.

First, experiments investigating joining and operating state are performed. The present analytical approaches will then be extended.

Table 1 shows the parameters that influence joining forces and maximal transmittable torsional load in the design of a KIF. The most important factors are the shaft diameter, the joining length, and the geometric interference (Figure 1).

**Table 1.** Parameters that influence knurled interference fits: ↗ represents a larger influence with an increasing parameter; ↘ represents an opposed influence with an increasing parameter, ( ) represents a hypothesized influence.

| Parameter (Increasing) | | Symbol | Influence on Maximal Joining Force | Influence on Maximal Torque |
|---|---|---|---|---|
| Shaft diameter | ↗ | $D_S$ | ↗ | ↗ |
| Hub outer-diameter | ↗ | $D_{oH}$ | (↗) | (↗) |
| Joining length | ↗ | $l_j$ | ↗ | ↗ |
| Shaft chamfer angle (SCA) | ↗ | $\phi$ | ↘ | ↘ |
| Geometric interference | ↗ | $I_{geo}$ | ↗ | ↗ |
| Hub-diameter ratio (HDR) | ↗ | $Q_H$ | (↘) | (↘) |

## 2. Materials and Methods

Investigations on hub outside-diameter are performed on specimens with flange geometries as shown in Figure 2. The figured hub specimens are manufactured from rods of AlMgSi1, EN AW-6082, with heat treatment T6. The shaft material is bearing steel 100Cr6. The basic shaft geometry is manufactured in the untreated state. Subsequently, the knurling is fabricated on the specimen by recursive axial forming. Finally, the specimen is hardened to a hardness of approximately 758 HV. The resulting hardness ratio of joined specimens is 1:7 (Table 2).

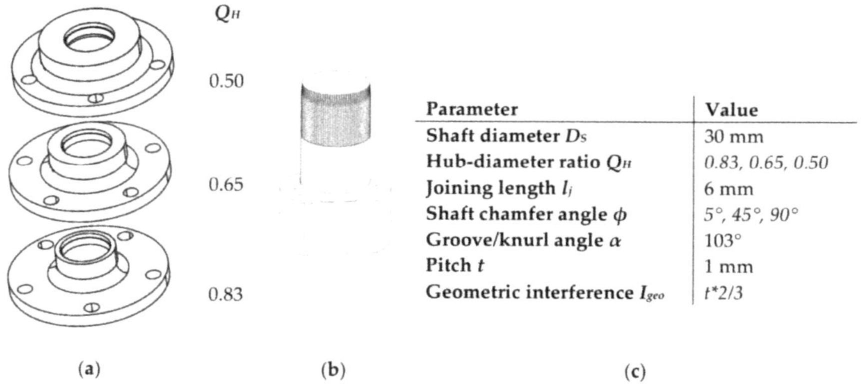


| Parameter | Value |
|---|---|
| **Shaft diameter** $D_S$ | 30 mm |
| **Hub-diameter ratio** $Q_H$ | *0.83, 0.65, 0.50* |
| **Joining length** $l_j$ | 6 mm |
| **Shaft chamfer angle** $\phi$ | *5°, 45°, 90°* |
| **Groove/knurl angle** $\alpha$ | 103° |
| **Pitch** $t$ | 1 mm |
| **Geometric interference** $I_{geo}$ | *t*2/3* |

(a)  (b)  (c)

**Figure 2.** (a) Aluminum hub, forging alloy AlMgSi1; (b) steel knurled shaft bearing steel 100Cr6; and (c) experimental parameters.

**Table 2.** Material properties.

| | Yield Stress $R_{p0.02}$ in MPa | Tensile Strength in MPa | Rupture Strain in % | Hardness |
|---|---|---|---|---|
| AlMgSi1 | 309 | 340 | 7.3 | 105 HV 10 |
| 100Cr6 | | 2525 * | | 758 HV 1 |

* calculated from hardness [6].

Figure 2 shows the specimen geometry as well as the chosen parameter values for the present study. The influence of different hub-diameter-ratios $Q_H$ on joining forces and maximal torque capacity differed by cutting and forming joining method is examined by varying the shaft chamfer angle $\phi$ for each HDR.

The joining process is performed in a special guiding appliance that guarantees the coaxial position of hub and shaft during the assembly in hydraulic press. The joining velocity $v_j$ was 0.5 mm/s due to comparability within the presented literature. The surfaces were cleaned before a dry joining process. The maximal joining force is derived from the gained force-stroke-signal.

After a one-day rest period the assembled connection is tested in a hydraulic torsional test bench, where the torque progression over the torque angle is measured until the failure of KIF occurs (stress-controlled test). At this value, the maximal torque capacity is obtained.

Material properties (Table 2) were estimated in tensile testing according to ISO 6892 [10] for the hub material AlMgSi1. In the case of the hardened steel shaft the hardness was measured according to ISO 6507 [11] and the tensile strength was calculated according to [6]. All tests were performed at room temperature.

## 3. Results

### 3.1. Experimental Results

Joining force plots shown in Figure 3a summarize the maximal forces of tested specimens. Every point represents a mean value of two repetitions with a maximal standard deviation of maximally 3.4% (joining force) and 2.9% (torque), respectively. At the lower HDRs of $Q_H = 0.65$ and $Q_H = 0.5$, the maximum joining force increases with decreasing SCA, peaking at $\alpha = 5°$. 12% lower values for $Q_H = 0.65$ are observed at this point.

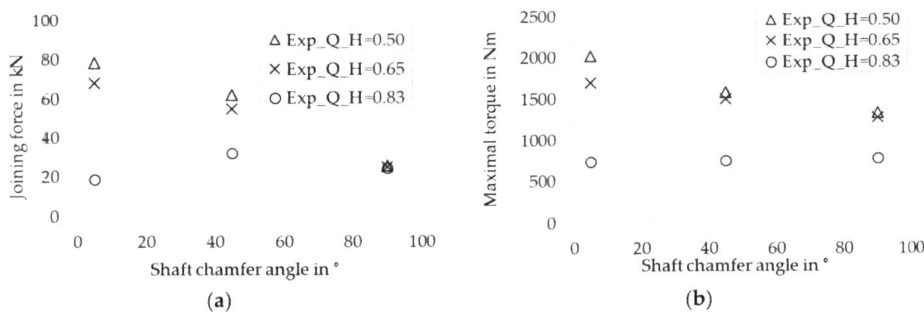


**Figure 3.** Results from hub-diameter ratio and shaft chamfer angle testing, (**a**) maximum joining force; and (**b**) maximal torque capacity.

In contrast, the thin walled hub with $Q_H = 0.83$ shows a plateau at lower joining forces. While all HDR forces are the same at $\alpha = 90°$, values diverge nonlinearly with decreasing SCA.

The forming process causes redirection of the axial force to a hub-expending radial force, which results in higher groove pressure and higher joining forces than the cutting process for hubs with higher wall thicknesses. In contrast, the thin walled hub does not show this effect due to high hub expansion during joining and the related lower plastic deformation and shorter height of teeth in the hub. The joining forces of forming KIF are about 20% lower than for cutting KIF.

Figure 3b shows the corresponding maximal torsional load, which is the load where the KIF fails. Similar trends of joining force (Figure 3a) can be observed. Except for the thin-walled hub, where the values at the plateau are lower, the forming connection also achieves a higher maximal torque. The higher torque capacity of forming KIF is rationalized by strain hardening of the ductile material AlMgSi1.

Additionally, the maximal torsional load of a thin walled KIF $Q_H = 0.83$ joined by cutting ($\alpha = 90°$) is about 40% lower than KIF with $Q_H = 0.5$ and $Q_H = 0.65$. This difference can be explained by radial expansion of the hub caused by the torque applied during testing. Enlargement of the hub inner diameter, caused by the force split of the torque on the sloping side of the formed/cut knurling, leads to decreased contact area. Interrelated reduction of the teeth shear area results in a smaller torque capacity for the hub with thin walls with $Q_H = 0.83$.

### 3.2. Analytical Approach for Calculation of Joining Force and Torque Capacity

The following description involves including the tested parameter in the available calculation method, thereby expanding the approach of Lätzer.

The first set of analysis highlights the impact of HDR on the formation of the counter profile in the hub. Figure 4 shows the different groove heights $h_{groove}$ dependency on SCA and HDR. The green area represents the neglected area in the field of thin hub and forming joining process. At this point, the contact area is reduced as a result of radial hub expansion.

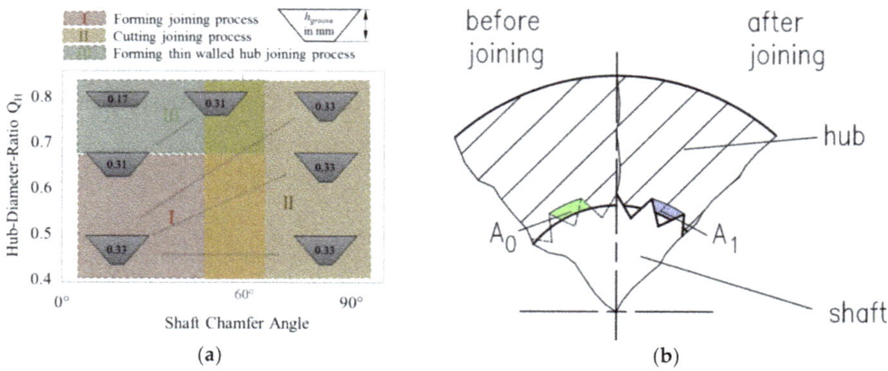


**Figure 4.** (**a**) Hub groove height formation after KIF joining; and (**b**) calculation areas for equivalent plastic strain $\varepsilon_{RPV}^{pl}$.

When low radial forces are initiated in the cutting process, groove formation is due to material removal resulting in equal groove heights (Figure 4a, yellow area).

Knurl forming on the thick hub, which is at lower HDR, leads to constant groove heights. In contrast to the cutting process, the material undergoes strain hardening, which is calculated in Equations (5) and (6) in the equivalent plastic strain term $\varepsilon_{KIF}^{pl}$.

The intersection of the described areas border the crossing parameter area, where the joining processes are mixed.

In addition to Lätzer's approach, the following equations are used for calculation of joining force $F_J$ and maximum torque $T_{torque,max}$ considering the tested parameters:

$$F_J = A_{contact} \cdot \mu \cdot p\left(\varepsilon_{KIF}^{pl}\right) \cdot K_J \tag{5}$$

$$T_{torque,max} = \left(\frac{D_S}{2} - h_{groove}(Q_H)\right) \cdot l_j \cdot i \cdot t \cdot \tau_S\left(\varepsilon_{KIF}^{pl}\right) \cdot K_T \tag{6}$$

Here the coefficient of friction $\mu$ is determined by a standardized tube friction method from [12]. The groove pressure $p\left(\varepsilon_{RPV}^{pl}\right)$ is calculated according to [6] to be equal to the flow stress $k_f\left(\varepsilon^{pl}\right)$.

The modification to the original Lätzer approach [6] consists of new reference areas $A_0$ and $A_1$ for calculation of equivalent plastic strain $\varepsilon_{RPV}^{pl}$, which is defined as:

$$\varepsilon_{KIF}^{pl} = \left|\ln\left(\frac{A_1}{A_0}\right)\right| \tag{7}$$

$$A_0 = t \cdot h_{groove}(Q_H) \tag{8}$$

$$A_1 = \frac{h_{knurl,shaft} \cdot t}{2} - \left(h_{knurl} - h_{groove}(Q_H)\right)^2 * \tan(\alpha/2) \tag{9}$$

The adjusted calculation of the areas $A_0$ and $A_1$ (Figure 4b) leads to a more accurate equivalent plastic strain which allows the numerical modelling of flow stress with the Ludwik approach [13]. With varying hub thickness, counter profile geometry also has to be considered. According to Figure 4, the required hub groove height $h_{groove}(Q_H)$ can be established by:

$$h_{groove}(Q_H) = \begin{cases} \frac{l_{geo}}{2}; & Q_H \leq 0.65; \; \forall \varphi \\ \frac{l_{geo}}{2} - \left(\frac{4}{5}Q_H + \frac{1}{2}\right) mm; & 0,85 > Q_H \geq 0.65; \; \varphi < 15° \\ \frac{l_{geo}}{2}; & 0,85 > Q_H > 0.55; \; \varphi > 60° \end{cases} \tag{10}$$

The height of the knurl $h_{knurl}$ results from the groove angle and the pitch:

$$h_{knurl} = \frac{t}{2 \cdot \tan(\alpha/2)} \tag{11}$$

Furthermore, the contact area $A_{contact}$ which influences the contact normal force is calculated by:

$$A_{contact} = i \cdot l_j \cdot \frac{2 \cdot h_{groove}(Q_H)}{\cos(\alpha/2)} \tag{12}$$

The influence of HDR and SCA is accounted for with the empirical-analytical functions $K_J(Q_H, \varphi)$ and $K_T(Q_H, \varphi)$, which include expansion of the hub and differentiate between the cutting and forming processes:

$$K_J(Q_H, \varphi) = C_{j1} + C_{j2} \cdot Q_H + C_{j3} \cdot \varphi + C_{j4} \cdot Q_H^2 + C_{j5} \cdot \varphi^2 + C_{j6} \cdot Q_H \cdot \varphi$$
$$\text{with } C_{j1} = 0.7759, C_{j2} = 1.6146, C_{j3} = -0.0102, \tag{13}$$
$$C_{j4} = -2,1717, C_{j5} == -0.0000145, C_{j6} = 0.00868$$

$$K_T(Q_H, \varphi) = \frac{C_{T1} + C_{T2} \cdot Q_H + C_{T3} \cdot \varphi}{1 + C_{T4} \cdot Q_H + C_{T5} \cdot Q_H^2 + C_{T6} \cdot \varphi}$$
$$\text{with } C_{T1} = 0.3017, C_{T2} = -0.1986, C_{T3} = 0.00158, \tag{14}$$
$$C_{T4} = -2.878, C_{T5} = 2.5396, C_{T6} = 0.003107$$

Figure 5 shows the cross-section curves for several HDR depending on SCA according to Equations (13) and (14). Corresponding to the experimental results, a decrease of load can be registered with decreasing SCA (cutting process) and increasing HDR (thin-walled hub).

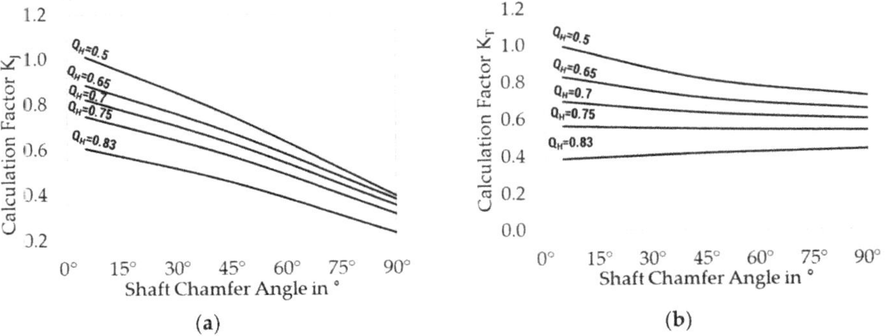


**Figure 5.** Factors influencing joining force (**a**) and maximal torque (**b**).

## 4. Discussion

As stated in the Introduction, the goal of the present research was to bridge knowledge gaps in the parameter range of the knurled interference fit. Joining force, as well as torsional load tests, were performed with varying hub-diameter ratios with the cutting and forming joining processes of shafts and hubs.

To conclude the experimental results, the present study supports previous research in this area. A similar influence of shaft chamfer angle was determined as found in [7,8]. As expected, including the new parameters in the known analytical approaches led to calculation errors, which overestimated the load capacity of KIF at higher HDR. Therefore, modifications of Lätzer's approach were made with the goal of minimizing calculation error. Considering the described points, Figure 6 shows a comparison of experimental results between "Lätzer-original" und "Lätzer-modified". Both show a low the deviation of calculated joining force of maximal 7% in case of forming joining (experiments no. 1–6, Figure 6a). For cutting connections ($\varphi = 90°$) deviations up to 38% of Lätzer-original" and 46% of "Lätzer-modified" are possibly caused by the unknown strain hardening formation on the knurl cutting tip during cutting process.

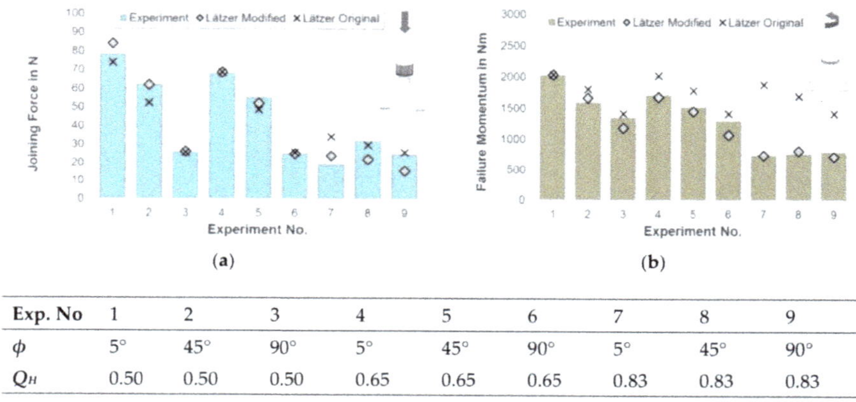


| Exp. No | 1 | 2 | 3 | 4 | 5 | 6 | 7 | 8 | 9 |
|---|---|---|---|---|---|---|---|---|---|
| $\phi$ | 5° | 45° | 90° | 5° | 45° | 90° | 5° | 45° | 90° |
| $Q_H$ | 0.50 | 0.50 | 0.50 | 0.65 | 0.65 | 0.65 | 0.83 | 0.83 | 0.83 |

**Figure 6.** Comparison of joining force (**a**) and torque (**b**) calculations to experiments, $D_S$ = 30 mm, $\alpha = 103°$, $I_{geo} = t \cdot \frac{2}{3}$, $t = 1$ mm, $\mu = 0.3$.

The accuracy of the new calculation approach is apparent with increasing HDR, where deviation of maximally 17% and $-3\%$ on average is reached (Figure 6b). In contrast, the "Lätzer-original" approach reaches deviations up to 155% (experiment no. 7). Acceptable values for maximal transmitting torque are achieved only for thick-walled hubs (experiment no. 1–3).

The described approach considers the specific specimen geometry. In practice different part geometries can lead to different load capacity. Therefore, a numerical approach for calculation of KIF is investigated additionally [14].

Furthermore, an axial load can be transmitted by the KIF connection due to remaining elastic portion of radial forming of the hub. The transmittable axial loading can be calculated regarding the axial strength described in [6]. The typical axial loads are about 30–45% of the joining force for cutting process and 60–90% of the joining force for forming process depending on the geometric interference of the KIF [6]. In dependency of rotational speed of the connection, the maximal transmittable force can differ according to [4].

## 5. Conclusions

Innovative design engineers often face novel problems, where missing knowledge leads to underachieving the structure potential. The knurl interference fit has a long record of use in different rotating applications, but the accurate calculation of possible transmittable torque is not fully known.

An example application can be found in modular structures where global structure should stay ductile (lower risk of crack) and contact parts should keep high hardness (wear-resistance). The connection of parts with these different properties are carried out by the KIF. As a result, increased functional integration of the part design is reached.

The results in present paper have been very promising for solving advanced design questions and minimizing experimental costs at the early construction phase of drive systems. Especially, the influence of the strain hardening, which corresponds to an improved material strength, is presently well described for aluminum hubs in dependence of different SCA and different HDR.

Future work concentrates on investigations with more material combinations, extending the analytic approach for more geometry variations. Additionally, a numerical model is built with the goal of investigating an expanded range of parameter. The validity of the present calculation approach for different knurled shapes, such as trapeze or inner knurling, is currently being investigated.

**Acknowledgments:** This research was made possible by a grant from the German Research Foundation DFG, for which are the authors very grateful.

**Author Contributions:** Lukas Suchy and Erhard Leidich conceived and designed the experiments; Lukas Suchy performed the experiments; Lukas Suchy and Thoralf Gerstmann analyzed the data; Thoralf Gerstmann and Birgit Awiszus contributed analysis tools; and Lukas Suchy wrote the paper.

**Conflicts of Interest:** The authors declare no conflict of interest.

## Abbreviations

| | |
|---|---|
| $A_{contact}$ | contact area |
| $A_0$ | reference undeformed area |
| $A_1$ | deformed area between knurls |
| $C_J$ | joining force correction factor |
| $C_T$ | torque correction factor |
| $D_S$ | Shaft diameter |
| $D_{iH}$ | Inner diameter of the hub |
| $D_{oH}$ | Outer diameter of the hub |
| $F_J$ | joining force of KIF |
| HDR | Hub-diameter-ratio |
| $h_{knurl,shaft}$ | shaft knurl height |
| $h_{groove}$ | formed groove height |
| $i$ | number of knurls |
| $K_{QH}$ | correction factor according to LÄTZER |
| $K_J$ | correction factor for joining force |
| $K_T$ | correction factor for maximal torque |
| KIF | knurled interference fit |
| $I_{geo}$ | geometric interference |
| $l_j$ | joining length |
| $p$ | contact pressure |
| SCR | shaft chamfer angle |
| $Q_H$ | hub-diameter-ratio |
| $t$ | pitch |
| $T_{\tau_s}$ | maximal transmittable torque |
| $\alpha$ | groove angle |
| $\varepsilon^{pl}_{KIF}$ | plastic strain |
| $\tau_S$ | shear blasting stress |
| $\varphi$ | shaft chamfer angle |

## References


1. Deutscher Institut für Normung. *DIN 7190-1. Pressverbände—Teil 1: Berechnungsgrundlagen und Gestaltungsregeln für Zylindrische Pressverbände*; Deutscher Institut für Normung e.V.: Berlin, Germany, 2017.
2. Wang, X.; Lou, Z.; Wang, X.; Xu, C. A new analytical method for press-fit curve prediction of interference fitting parts. *J. Mater. Process. Technol.* **2017**, *250*, 16–24. [CrossRef]
3. Hava, H.; Music, O.; Koç, A.; Karadogan, C.; Bayram, Ç. Analysis of elastic-plastic interference-fit joints. *Procedia Eng.* **2014**, *81*, 2030–2035.
4. Kovan, V. Separation frequency analysis of interference fitted hollow shaft–hub connections by finite element method. *Adv. Eng. Softw.* **2011**, *42*, 644–648. [CrossRef]
5. Deutscher Institut für Normung. *DIN 82. Rändel*; Deutscher Institut für Normung e.V.: Berlin, Germany, 1973.
6. Lätzer, M. Untersuchungen zum Füge- und Übertragungsverhalten Torsionsbelasteter Stahl-Aluminium-Rändelpressverbindungen. Ph.D. Thesis, Technische Universität Chemnitz, Chemnitz, Germany, 2015.
7. Lätzer, M.; Leidich, E.; Kleditzsch, S.; Awiszus, B. Untersuchungen zum Übertragungsverhalten von Rändelpressverbänden aus Stahl-Aluminium. In *Untersuchungen zum Übertragungsverhalten von Rändelpressverbänden aus Stahl-Aluminium*; Forschung im Ingenieurwesen; Springer: Berlin, Germany, 2015; Volume 79, pp. 41–56.
8. Bader, M. Das Übertragungsverhalten von Pressverbänden und die Daraus Abgeleitete Optimierung Einer Formschlüssigen Welle-Nabe-Verbindung. Ph.D. Dissertation, Technische Universität Graz, Graz, Austria, 2009.
9. Hirota, K.; Kitamura, K.; Ukai, Y.; Matsunaga, K. Mechanical joining of shaft and holed disc in rotational and axial directions. *Procedia Eng.* **2017**, *207*, 980–985. [CrossRef]
10. International Organization for Standardization. *ISO 6892-1. Metallic Materials—Tensile Testing—Part 1: Method of Test at Room Temperature*; ISO: Geneva, Switzerland, 2017.
11. International Organization for Standardization. *ISO 6507-1. Metallic Materials—Vickers Hardness Test—Part 1: Test Method*; ISO: Geneva, Switzerland, 2005.
12. Reiß, F.; Gräfensteiner, M. *Transferability of Static Friction Coefficients*; FVV Final Report 1148; Forschungsvereinigung Verbrennungsmotoren e.V.: Chemnitz, Germany, 2018.
13. Ludwik, P. *Elemente der Technologischen Mechanik*; Springer: Berlin, Germany, 1909.
14. Gerstmann, T.; Awiszus, B.; Suchý, L.; Leidich, E. Numerische Analyse des Montage- und Übertragungsverhaltens von Rändelpressverbänden. In *VDI-Berichte 2287*; VDI Wissensforum GmbH: Karlsruhe, Germany, 2016.

*Article*

# Theoretical and Experimental Studies of Over-Polishing of Silicon Carbide in Annular Polishing

**Junjie Zhang** [1,*]**, La Han** [1]**, Haiying Liu** [2]**, Yikai Shi** [3]**, Yongda Yan** [1] **and Tao Sun** [1]

[1]  Center for Precision Engineering, Harbin Institute of Technology, Harbin 150001, China;
    m13091711685@163.com (L.H.); yanyongda@hit.edu.cn (Y.Y.); spm@hit.edu.cn (T.S.)
[2]  Xiaguang Optical Electron Co., Ltd, Yangzhou 225127, China; xiaohaibao@sina.com
[3]  School of Mechanical and Electrical Engineering and Automation, National University
    of Defense Technology, Changsha 410073, China; sykai1995@163.com
*  Correspondence: zhjj505@gmail.com

Received: 7 February 2018; Accepted: 3 April 2018; Published: 4 April 2018

**Abstract:** Annular polishing technology is an important optical machining method for achieving a high-precision mirror surface on silicon carbide. However, the inevitable over-polishing of the specimen edge in annular polishing deteriorates achieved surface quality. In the present work, we first analytically investigate the kinematic coupling of multiple relative motions in the annular polishing process and subsequently derive an analytical model that addresses the principle of material removal at specimen edge based on the Preston equation and the rigid body contact model. We then perform finite element simulations and experiments involving annular polishing of silicon carbide (SiC), which jointly exhibit agreement with the derived analytical model of material removal.

**Keywords:** silicon carbide; annular polishing; material removal; over-polishing; finite element

## 1. Introduction

Silicon carbide (SiC) is one of the preferred materials for manufacturing optical mirrors due to its unique characteristics of low density, high strength, low thermal expansion, high thermal conductivity and high chemical inertness [1,2]. According to the theory of total integrated scattering (TIS), the total surface scattering of a mirror is closely related to its surface roughness. Specifically, with the increase of surface roughness, the TIS rises sharply in conjunction with the decrease of reflectivity, which accordingly results in the degradation of the imaging quality of the optical system [3]. Therefore, achieving an ultra-smooth surface by optical machining methods is critical for the performance of SiC mirrors. At present, annular polishing technology is one of the main methods that has been widely used to obtain high-precision SiC mirrors [4–7].

The inevitable over-polishing of specimen edge that occurs during annular polishing significantly deteriorates machined surface qualities, such as flatness. Additionally, a fundamental understanding of material removal is required to minimize over-polishing. Material removal during the annular polishing process is complex due to the kinematic coupling of multiple relative motions between the specimen and polishing disc. At present, the Preston equation is widely used to describe the material removal process in annular polishing. For instance, Ji et al. [8] investigated the relationship between surface quality and polishing condition and consequently established the trajectory equation for a point on the single-side polishing part relative to the polishing pad by the coordinate transformation method. Wenski et al. [9] obtained improved non-uniformity for material removal at specimen edge by optimizing the trajectory of the polishing machine. However, the Preston equation still needed to be refined. Nanz et al. [10,11] found, through polishing experiments, that although the product of surface pressure of the specimen and its rotate speed is zero, the polishing removal rate is not zero due to chemistry action. Consequently, a compensation parameter must be introduced to correct underestimation by the Preston equation and which fits well with specific values in the experiments. Luo et al. [12] modified the Preston equation according to the results of both experiments and theoretical calculations.

Although previous analytical and experimental studies have qualitatively analyzed average material removal in the annular polishing process, there is limited work focusing on the over-polishing of specimen edge. In particular, there is no model of a material removal equation that takes into account the phenomenon of edge over-polishing, as the Preston equation only considers the average pressure of the polishing process. Therefore, in the present work, we first analyze the kinematic coupling of multiple relative motions between the specimen and polishing disc in the annular polishing process. Subsequently, based on the Preston equation and the rigid body contact model, we derive an analytical model that addresses the principle of material removal at specimen edge. Finally, finite element (FE) simulations and experiments of annular polishing of SiC are performed to evaluate the accuracy and efficiency of the derived analytical model.

## 2. Analytical Investigation of Annular Polishing

### 2.1. Kinematic Coupling of Motions

Figure 1a illustrates a typical annular polisher, which consists of a polishing disc, a carrier disc and a swinging bracket. The specimen is pasted on the carrier disc with paraffin, which means that the specimen has a synchronous speed with the carrier disc. The applied pressure is provided by the weight of the carrier disc. Accordingly, Figure 1b illustrates the simplified motion diagram of the annular polishing, which indicates that the kinematic coupling of relative motions includes the rotation of the polishing disc and the rotation of the carrier disc. We note that the swinging of the carrier disc leads to back and forth movements by the specimen, which consequently reduce the propensity of over-polishing of specimen edge. Therefore, to magnify the over-polishing phenomenon, in the present work the back and forth swinging motions are not considered in the kinematic coupling. As indicated in Figure 1b, the speed of the carrier disc is $\omega$ and the speed of the polishing disc is $\delta$. The distance from the center of specimen $O_2$ to point A on the carrier disc is $r$ and the distance from the center of the polishing disc $O_1$ to the center of carrier disc $O_2$ is $R$. The angle between the line segments $AO_2$ and $O_1O_2$ is $\theta$. The speed at point A on the polishing disc relative to $O_1$ is $V_1$ and the speed at point A on the specimen relative to $O_2$ is $V_2$, so the relative speed of the specimen and the polishing disc at point A is $V$, which can be derived from Equation (1) [13]:

$$V(r,\theta) = [R^2\delta^2 + r^2(\delta - \omega)^2 + 2rR\delta(\delta - \omega)\cos\theta]^{1/2} \tag{1}$$

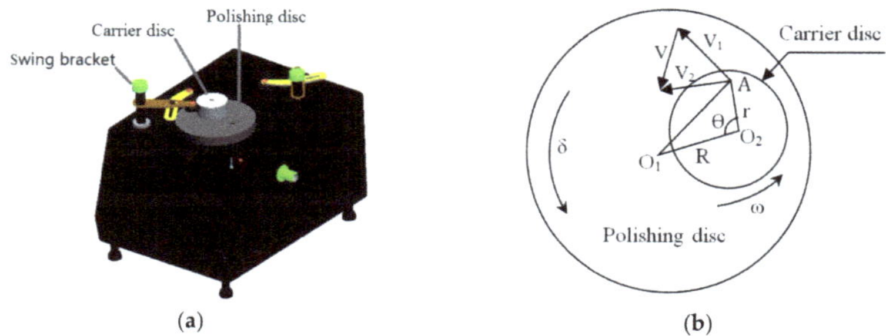


**Figure 1.** Illustration of annular polishing. (**a**) Model of annular polisher; (**b**) motion analysis of annular polishing.

### 2.2. Preston Equation

Both material removal rate and surface quality of the specimen in the annular polishing process are strongly affected by processing parameters, which have complex interactions. Preston et al. simplified the Preston equation to characterize the relationship between material removal and polishing speed $V$, applied pressure $P$, and other external factors, as shown in Equation (2) [14]:

$$\frac{dh}{dt} = kPV = kP\frac{ds}{dt} \tag{2}$$

where $h$ is the amount of material removal and $k$ is a proportional constant that is related to various environmental factors. Therefore, the amount of material removal at one specific point can be derived according to Equation (2) [15,16]. However, contact between the specimen and polishing disc changes dynamically with polishing time, which induces uncertainties in analytical investigation of the annular polishing process. Therefore, three main assumptions are made to simplify the operation: (1) The specimen and polishing disc are completely in contact without separation; (2) the applied pressure does not change over polishing time; and (3) the proportional constant $k$ does not change over polishing time.

According to the Preston equation, the amount of material removal within a specific polishing time can be derived by integrating time $T$, as shown in Equation (3):

$$h(r) = k \int_0^T P(r,t) \cdot V(r,t) dt \tag{3}$$

It can be seen from Equation (3) that material removal is only dependent on the resultant speed $V$, given that $k$ and $P$ are constant. By substituting the relative velocity derived from Equation (1) into Equation (3), material removal at point $A$ can be derived, as shown in Equation (4):

$$h(r) = k \int_0^T P(r,t) \cdot [R^2\delta^2 + r^2(\delta - \omega)^2 + 2rR\delta(\delta - \omega)\cos\theta]^{1/2} dt \tag{4}$$

### 2.3. Rigid Body Contact Model

It can be seen in Equation (4) that the pressure of each point on the specimen must be accounted for precisely when deriving material removal. Contact between the specimen and polishing disc in annular polishing can be analyzed by using the rigid body contact model, as illustrated in Figure 2 [17].

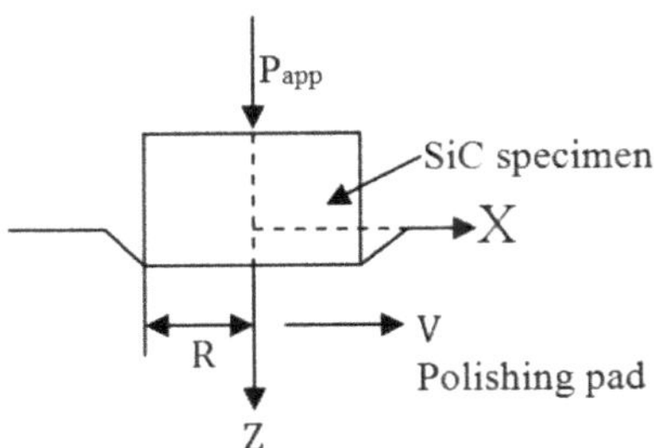


**Figure 2.** Illustration of the rigid body contact model.

The rigid body contact model, as a two-dimensional plane strain model, can be used to analyze the distribution of contact pressure on the contact surface between the specimen and polishing disc. Considering the much higher stiffness (Mohs hardness of 9.5 and elastic modulus of 360 GPa) of SiC compared to that of a polishing disc made of cast iron (Mohs hardness of 4.5 and elastic modulus of 170 GPa), the SiC specimen is treated as a rigid body and the polishing disc is treated as an elastic material with small elastic modulus (elastic modulus of 3 MPa). Therefore, the shape of the SiC specimen does not change after being pressed into the polishing disc. Furthermore, it is assumed that the deformation of the polishing disc does not affect the result. The distribution of contact pressure on the specimen surface is expressed in Equations (5) or (6):

$$P(x) = \frac{P_{app}}{\pi(R^2 - x^2)^{1/2}} \tag{5}$$

$$P(x) = \frac{P_{app}\cos(\pi\gamma)}{\pi(R^2 - x^2)^{1/2}}\left(\frac{R+x}{R-x}\right)^{\gamma} \tag{6}$$

where $P_{app}$ is the pressure applied to the upper surface of specimen and $R$ is the radius of specimen. $\gamma$ can be expressed by:

$$\cot\pi\gamma = -\frac{2(1-v)}{f(1-2v)} \tag{7}$$

where $f$ is the friction coefficient between the specimen and the polishing pad and $v$ is the Poisson's ratio of the polishing pad. While Equation (5) does not consider the friction between contacting surfaces, Equation (6) provides a more accurate description of the contact by considering the frictional force between the two surfaces. By substituting Equation (6) into Equation (4), the removal equation for the specimen in annular polishing can be derived, as seen in Equation (8):

$$h(r) = k\int_0^T \frac{P_{app}\cos(\pi\gamma)}{\pi(R^2 - x^2)^{1/2}}\left(\frac{R+x}{R-x}\right)^{\gamma} \cdot [R^2\delta^2 + r^2(\delta-\omega)^2 + 2rR\delta(\delta-\omega)\cos\theta]^{1/2} dt \tag{8}$$

It can be seen from Equations (5) and (6) that contact pressure near specimen edge (i.e., $r$ approaching $R$) increases sharply. Therefore, Equation (8) also applies to the qualitative description of material removal when considering over-polishing of specimen edge.

## 3. FE Simulation of Annular Polishing

### 3.1. FE Modeling

To verify the accuracy of the derived model of material removal presented in Equation (8), FE simulations of annular polishing of SiC were performed by using ABAQUS software, with an emphasis on the distribution of contact pressure on specimen surface. Similarly to the analytical investigation, in the FE simulations the motion of carrier swing was not considered. Accordingly, the swing bracket was omitted in the FE model. Figure 3 shows that the FE model of annular polishing consisted of a SiC specimen colored in blue, a polishing pad in red and a polishing disc in gray, which were treated as deformable parts. The upper surface of the polishing disc and the lower surface of the polishing pad were bound entirely. The polishing disc had a synchronous speed with the polishing pad. Table 1 lists material properties of the three parts. Furthermore, it was assumed that the properties of the SiC specimen were isotropic.

**Table 1.** Material parameters of parts.

|  | Elastic Modulus/Mpa | Poisson's Ratio | Density Kg/m$^3$ |
|---|---|---|---|
| Silicon carbide (SiC) specimen | 362,390 | 0.163 | 3080 |
| Polishing pad | 3 | 0.1 | 260 |
| Polishing disc | 148,000 | 0.31 | 7200 |

It is known that grid configuration greatly affects the accuracy of prediction results in FE simulations. The three aspects of cell type, cell shape and grid density need to be considered when configuring grids. In the present work, a structured mesh was used to divide meshes by using a linear hexahedral C3D8I element with eight nodes. Grid density increased with increasing distance from the center of specimen.

Furthermore, the mesh of the polishing pad was denser than that of the polishing disc, as shown in Figure 3.

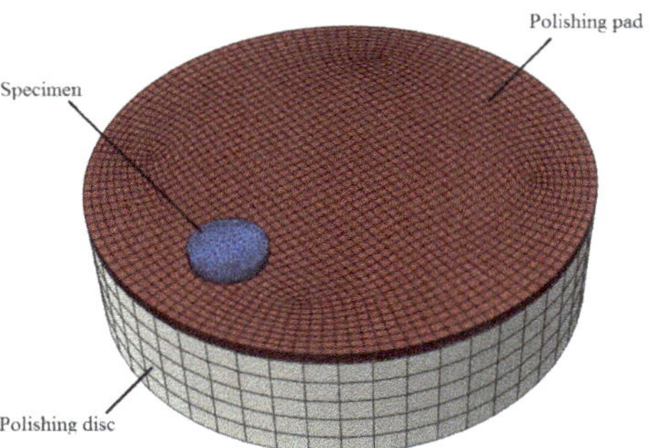


**Figure 3.** Finite element model of annular polishing of silicon carbide (SiC).

In the FE simulation, an explicit dynamical analysis step with a total time of 3 s was used. The selection of 3 s was to ensure that the specimen rotated more than two revolutions, which enabled us to obtain more accurate polishing results. It should also be noted that the time cannot be too long due to high computational costs and minimal change in results. The contact between the specimen and polishing pad was set as a general contact with a friction coefficient of 0.1. A uniform pressure of 0.068 MPa was imposed on the specimen.

*3.2. Simulation Results and Discussion*

Figure 4 presents the distribution of contact pressure and deformation of the SiC specimen surface after FE simulation of annular polishing, in which the polishing disc and specimen have the same speed of 100 r/min. In Figure 4a, distributions of contact pressure in three random radial directions highlighted by the three randomly selected red dash lines are analyzed. For each direction, 30 uniformly-spaced points along the red lines are selected. Accordingly, Figure 4c plots variations of contact pressure along the three directions as a function of off-center distance, which shows that the contact pressure in the vicinity of the center of the specimen is distributed uniformly. However, when the off-center distance is higher than 10 mm, the contact pressure gradually increases with increasing off-center distance, and finally reaches the maximum value at the specimen edge. Figure 4b shows the contour of normal strain after polishing, which is related to the material removal of specimen surface. It can be seen in Figure 4b that the material removal at specimen edge is significantly higher than in the middle of the specimen.

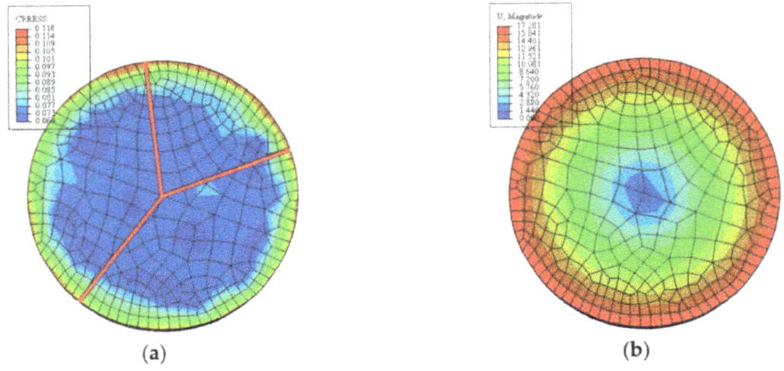

(a)  (b)

**Figure 4.** *Cont.*

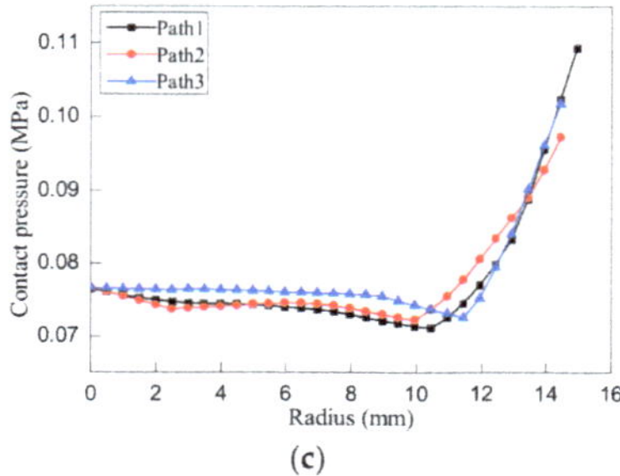

(c)

**Figure 4.** Distributions of contact pressure and normal strain in finite element (FE) simulations of annular polishing of SiC. (**a**) Contour of contact pressure; (**b**) contour of normal strain; and (**c**) variations of contact pressures.

We further evaluated the influence of the typical parameters (speed, Poisson's ratio and elastic modulus of polishing pad) on the distribution of contact pressure. Figure 5 plots variations of contact pressure along three directions as a function of off-center distance under different parameters. It was found, as shown in Figure 5, that for each parameter, the contact pressure of the specimen showed similar trends, first remaining stable and then increasing sharply when near the edge. The above results verify the pressure distribution equation obtained by the rigid body contact model presented in Equations (5), (6) and (8).

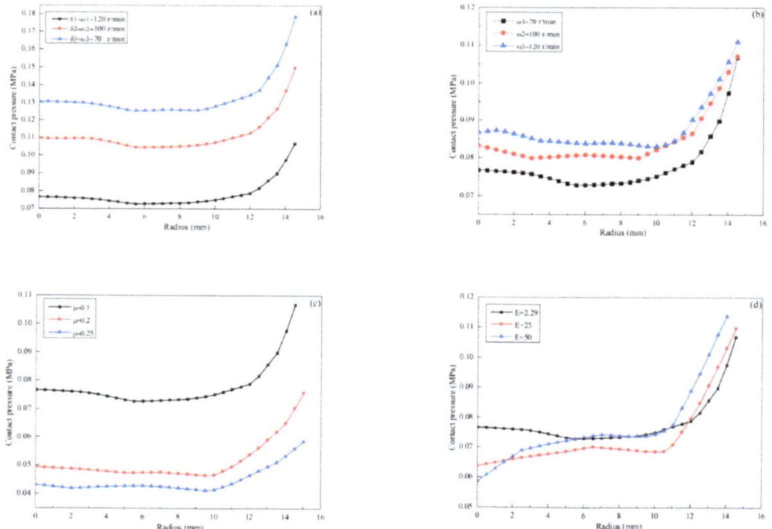

**Figure 5.** Distribution of contact pressure on the SiC specimen surface at different parameters. (**a**) Speed (specimen and polishing plate rotate at equal speed); (**b**) speed (keeping the speed of the polishing disc as 100 r/min, changing the speed of the specimen); (**c**) Poisson's ratio of polishing pad; and (**d**) elastic modulus of polishing pad.

## 4. Experimental Study of Annular Polishing of SiC

In addition to the FE simulations that verified the occurrence of over-polishing of specimen edge, annular polishing experiments of SiC were also carried out to further verify the derived model of material removal presented in Equation (8). All the annular polishing experiments were performed by using the UNIPOL-802 precision auto lapping and polishing machine, as shown in Figure 6. The specific parameters used in the annular polishing experiments were: That the velocities of the polyurethane polishing pad and specimen were the same at 100 r/min, the polishing time was 60 min, and a diamond suspension solution was used. After polishing, the obtained surface was measured by using a surface profiler.

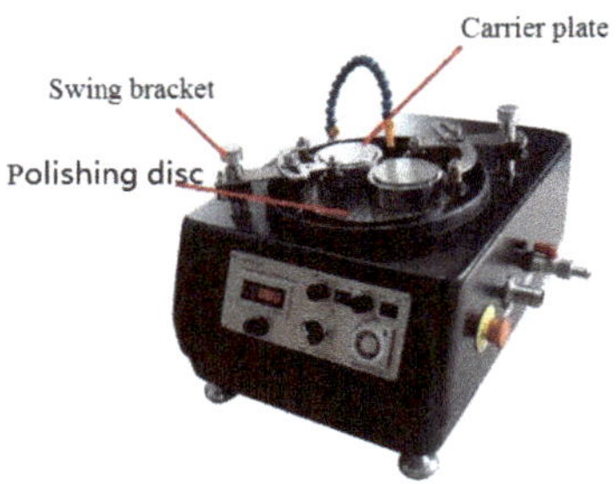


**Figure 6.** Precision auto lapping and polishing machine.

Figure 7 plots the variation of measured relative surface height with off-center distance. The relative surface height is defined as the change in measured surface height after polishing with respect to surface height before polishing. In order to show the mutual verification of the experimental results and the FE simulation results, Figure 7 further plots the variations of contact pressure shown in Figure 4. It can be seen in Figure 7 that within the off-center distance ranging from 0 to 13 mm, the relative surface height of the specimen remains relatively unchanged. Upon a further increase of off-center distance from 13 to 15 mm, which approaches to the specimen edge, however, the relative surface height of the specimen dropped sharply. This indicates that the amount of material removal also increased sharply. Figure 7 also demonstrates that the experimental results are in agreement with the FE simulation results, as material removal during polishing is proportional to contact pressure. Specifically, there is sharp increase in both pressure and material removal at specimen edge, indicating the occurrence of an over-polishing phenomenon. Therefore, it is indicated by both the simulation and experimental results that the derived analytical model of material removal presented in Equation (8) is indeed capable of describing material removal particularly related to the over-polishing of specimen edge during annular polishing of SiC.

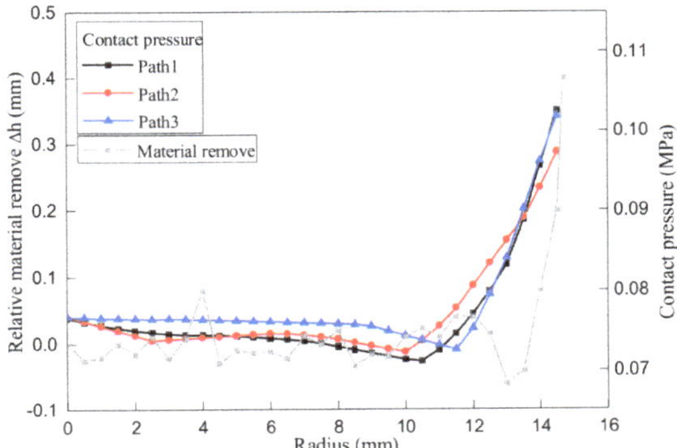


**Figure 7.** Distributions of material removal by experiment and contact pressure by FE simulation.

## 5. Conclusions

In the present work, we performed an analytical investigation, FE simulation and an experimental study to investigate the fundamentals of material removal during annular polishing of SiC, with an emphasis on the underlying mechanisms of over-polishing of specimen edge. According to the analytical investigations of the kinematic coupling of multiple relative motions between the specimen and polishing disc, an analytical model of material removal during the annular polishing process that further accounts for the over-polishing of specimen edge was derived based on the Preston equation and the rigid body contact model. Subsequent FE modeling and simulation showed the non-uniform distribution of contact pressure and the normal strain of the specimen in the annular polishing, i.e., the occurrence of over-polishing at specimen edge, which was further verified by the polishing experiments.

**Acknowledgments:** The authors acknowledge financial support from the National Natural Science Foundation of China (NSFC) and the German Research Foundation (DFG) International Joint Research Program (51761135106).

**Author Contributions:** J.Z., L.H. and T.S. conceived and designed the experiments; L.H. performed the analytical investigation and FE simulations, Y.S. performed the experiments; J.Z., L.H. and Y.Y. analyzed the data; J.Z. and L.H. wrote the paper.

**Conflicts of Interest:** The authors declare no conflict of interest.

## References

1.  Robichaud, J.L.; Schwartz, J.; Landry, D.; Glenn, W.; Rider, B.; Chung, M. Recent advances in reaction bonded silicon carbide optics and optical systems. *SPIE* **2005**, *5868*, 586802.
2.  Wang, X.C.; Wang, C.C.; Shen, X.T.; Sun, F.H. Potential material for fabricating optical mirrors: Polished diamond coated silicon carbide. *Appl. Opt.* **2017**, *56*, 4113–4122. [CrossRef] [PubMed]
3.  Cook, F.; Brown, N.; Prochnow, E. Annular lapping of precision optical flatware. *Opt. Eng.* **1976**, *15*, 450–458. [CrossRef]
4.  Gritti, F.; Guiochon, G. Gradient chromatography under constant frictional heat: Realization and application. *J. Chromatogr. A* **2013**, *1289*, 1–12. [CrossRef] [PubMed]
5.  Hutchings, I.M.; Xu, Y.; Sanchez, E.; Ibanez, M.J.; Quereda, M.F. Development of surface finish during the polishing of porcelain ceramic tiles. *J. Mater. Sci.* **2005**, *40*, 37–42. [CrossRef]
6.  Fan, Q.T.; Zhu, J.Q.; Zhang, B.A. Effect of the geometry of workpiece on polishing velocity in free annular polishing. *Chin. Opt. Lett.* **2007**, *5*, 298–300.
7.  Hashimoto, Y.; Oshika, S.; Suzuki, N.; Shamoto, E. A new contact model of pad surface asperities utilizing measured geometrical features. In Proceedings of the International Conference on Planarization/CMP Technology (ICPT), Chandler, AZ, USA, 30 September–2 October 2015; pp. 1–4.
8.  Yongji, J.I.N. Study on Mechanism of CMP Polishing Motion. *Equip. Electron. Prod. Manuf.* **2005**, *34*, 37–41.
9.  Wenski, G.; Altmann, T.; Winkler, W.; Heier, G.; Hölker, G. Doubleside polishing—A technology mandatory for 300 mm wafer manufacturing. *Mater. Sci. Semicond. Process.* **2003**, *5*, 375–380. [CrossRef]
10. Nanz, G.; Lawrence, E.C. Modeling of chemical mechanical polishing: A review. *IEEE Trans. Semicond. Manuf.* **1995**, *8*, 382–389. [CrossRef]
11. Qin, K.D.; Moudgil, B.; Park, C.W. A chemical mechanical polishing model incorporating both the chemical and mechanical effects. *Thin Solid Film* **2004**, *446*, 227–286. [CrossRef]
12. Luo, Q.; Ramarajan, S.; Babuu, S.V. Modification of the Preston equation for the chemical mechanical polishing of copper. *Thin Solid Film* **1998**, *335*, 160–167. [CrossRef]
13. Goedecke, A.; Jackson, R.L.; Mock, R. A fractal expansion of a three dimensional elastic-plastic multi-scale rough surface contact model. *Tribol. Int.* **2013**, *59*, 230–239. [CrossRef]
14. Preston, F. The theory and design of plate glass polishing machines. *J. Soc. Glass Technol.* **1927**, *11*, 214–256.
15. Li, S.X.; Gang, Y.; Zhang, J.C. Single-row laser beam with energy strengthened ends for annular scanning laser surface hardening of largemetal components. *Sci. Chin. Phys. Mech. Astron.* **2013**, *56*, 1074–1078. [CrossRef]

16. Zhao, D.L.; Ma, X.; Liu, B.; Xie, L. Rainfall effect on wind waves and the turbulence beneath air-sea interface. *Acta Oceanol. Sin.* **2013**, *32*, 10–20. [CrossRef]
17. Johnson, K.L. *Contact Mechanics*; Cambridge University Press: Cambridge, UK, 1985; pp. 44–48.

*Article*

# Effects of Setting Errors (Insert Run-Outs) on Surface Roughness in Face Milling When Using Circular Inserts

**Csaba Felhő * and János Kundrák**

Institute of Manufacturing Science, University of Miskolc, Egyetemvaros, 3515 Miskolc, Hungary;
janos.kundrak@uni-miskolc.hu
* Correspondence: csaba.felho@uni-miskolc.hu; Tel.: +36-46-565-160

Received: 6 March 2018; Accepted: 27 March 2018; Published: 2 April 2018

**Abstract:** In face milling, the roughness of the machined surface varies due to the movement of the cutting edge. Changes in roughness parameter values in the axis of rotation (symmetry plane) have been examined at a constant depth of cut for symmetrical milling. In this paper, the effect of increasing feed per tooth on the topography of the surface is studied in fly-cutting and in multi-point face milling. The study takes into account the axial run-out of the inserts. Theoretical roughness values were modelled, the real values were tested in experiments and in both cases the impact of the run-out of the cutting edges and the change of the chip cross-section were also taken into account. Based on the performed experiments it can be stated that the accuracy of the introduced roughness prediction method increases with the increase in feed and therefore the application of the method in the case of high-feed milling is particularly effective. The results have also shown that the run-out of the insert significantly effects the roughness of the milled surfaces and therefore the measurement and minimization of these setting errors is essential.

**Keywords:** face milling; surface roughness; feed; insert run-out

## 1. Introduction

Increasing the productivity of machining requires more intensive material separation. This can be characterized by increasing both the values of the cutting data (the removed chip cross-sections) and the Material Removal Rate (MRR) and the Surface Rate (SR) values. At the same time, consideration should be given to the rigidity of the Machine/Fixture/Tool/Workpiece (MFTW) system and to the required/expected quality of the machined surface. Among the cutting data, increasing the feed directly affects the material removal rate (MRR) but due to its increased value, the roughness of the machined surfaces may deteriorate, since it primarily determines the height of the roughness peaks on the surface. Therefore, the experimental examination and estimation of the expected roughness characteristics for the machined surface, which may limit the applicable feed values, becomes increasingly important. The relevance of this topic is demonstrated by the fact that many researchers are concerned with the investigation of the roughness of milled surfaces, using different approaches and test methods.

The simplest machining process for modelling surface roughness is turning. Numerous studies deal with the simulation of theoretical roughness in turning, the most recent ones being briefly presented here. He et al. [1] introduced a systematic review of influencing factors and theoretical modelling methods of surface roughness in turning process. They found that the most important factors are the kinematic and dynamic properties of the machine tool, the geometrical parameters of the cutting tool, the properties of the workpiece material and the applied coolant. They have classified these factors as easy and difficult to modelling and the coolant and workpiece material properties belong to the second group. Tomov et al. [2] presented a methodology for modelling and predicting the roughness shape in longitudinal turning that is utilized for both the kinematical-geometrical replication of the cutting tool geometry onto the machined surface and other cutting conditions and factors that are considered a black box—the latter include mechanical properties, thermal preparation, the material of the inserts, the positioning of the tool, the working conditions of the machine, the cutting force, the cutting temperatures, the tool wear, the vibrations of the workpiece and so forth. In the case of milling, it is more difficult to precisely map the theoretical topography due to the multi-point tool design and to the more complex kinematic conditions. Nevertheless, there are many studies on this topic too, some of which are described here in brief. Grzenda and Bustillo [3] proposed a hybrid algorithm which combines a Genetic Algorithm (GA) with Artificial Neural Networks (ANN) for the selection of major parameters for the prediction of surface roughness in high-torque face milling operations. The input data set includes the following parameters: cutting tool geometry, technological parameters and cutting phenomena. Colak et al. [4] used a Gene Expression Programming (GEP) method to predict the surface roughness of end-milled surfaces from cutting parameters such as cutting speed, feed and depth of cut. El-Sonbaty et al. [5] used Artificial Neural Network (ANN) models to analyse and predict the relationship between the cutting conditions and the corresponding fractal parameters of machined surfaces in face milling. The input parameters of the ANNs were the following: rotational speed (n), feed (f), depth of cut ($a_p$), pre-tool flank wear and vibration level. The output parameters were the corresponding calculated fractal parameters: fractal dimension "D" and vertical scaling parameter "G." Tseng et al. [6] used design of experiments (DoE) to determine the significant factors and then fuzzy logic approach for the prediction of surface roughness. The factors considered for DoE were the depth of cut ($a_p$), feed per tooth ($f_z$), cutting speed ($v_c$), tool nose radius ($r_\varepsilon$), the use of cutting fluid and the three components of the cutting force (Fx, Fy, Fz). The impact of the most important factors on the surface roughness in semi-solid AA 7075 face milling were investigated in Reference [7]. The considered factors were the spindle speed (n), feed rate ($v_f$) and depth of cut ($a_p$) and it was found that the surface roughness was mostly affected by the feed rate ratio and the speed, while the impact of the depth of cut was insignificant. A surface reconstruction model is introduced in Reference [8] that is based on a methodology developed for the prediction of cutting forces in freeform milling. From the global and local geometry of the tool, initial surface and tool path, this approach allows the prediction of cutting forces, surface form and roughness directly from CAM data. A generalized mathematical model of roughness formation is introduced in Reference [9] for surfaces generated by round-nose tools. The model enables the calculation of surface roughness taking into account the tool characteristics, undeformed chip thickness, tool vibrations, tool runout (for multi-point tools) and tool wear. The developed mathematical model was verified by surfaces sculptured by face milling. A statistical model is presented in Reference [10] for surface roughness estimation in high-speed flat end milling using machining variables such as spindle speed (n), feed rate ($v_f$), depth of cut ($a_p$) and width of cut ($a_e$). In Reference [11] optimal cutting parameters were determined, resulting in minimal surface roughness in up peripheral milling of Ti-6Al-4V alloys. Theoretical roughness indexes (Ra) were determined by a model utilizing an artificial neural network. RSM and ANOVA were used to determine the input and output parameters. Miko et al. [12] used an experimental verification method to analyse the of cusp height evolution in end ball milling. The relationship which describe the effect of the direction of milling cutter motion on cusp heights has been derived from the geometrical interpretation of inclined elementary surface.

Shyha et al. [13] studied the microstructure of machined surfaces and chip formation in step-shoulder down-milling, using Ti-6Al-4V material with water-miscible vegetable oil-based coolant and lubricant. It has been found that the micro-geometry of machined surfaces largely depends on the cutting speed and the flow rate of the cutting fluid. Kilickap et al. [14] investigated the effect of cutting speed, feed rate and depth of cut on cutting force, surface roughness and tool wear when milling Ti-6242S alloy. The experimental data were compared with values determined by ANN and RSM methods.

A so-called desirability approach was applied in Reference [15] for the modelling of the following output responses by Response Surface Methodology (RMS): surface roughness (Ra), cutting force (Fc), cutting power (Pc), specific cutting force (Ks) and metal removal rate (MRR), during the face milling of the austenitic stainless steel X2CrNi18-9 with coated carbide inserts (GC4040). A full factorial design (L27) is selected for the experiments and ANOVA is used in order to evaluate the influence of the cutting parameters of cutting speed ($v_c$), feed per tooth and depth of cut ($a_p$) on the out-put responses. Nguyen and Hsu [16] investigated the effect of the cutting parameters on the surface roughness parameter Ra with a combination of the Taguchi method and the RSM. They have used quadratic mathematical modelling to estimate and the desirability function to minimize the Ra parameter. The effects of the insert runout errors and the variation of the feed rate on the surface roughness and the dimensional accuracy were analysed in Reference [17] in a face-milling operation using a surface roughness model. Schmitz et al. [18] investigated the effect of milling cutter teeth runout on surface topography, surface location error and stability in end milling. They pointed out that the cutter runout is an important issue in machining as commercially-available cutter bodies often exhibit significant deviations in milling insert locations in axial and radial directions; therefore, the chip load on the individual cutting teeth varies periodically. A numerical calculation model is presented in Reference [19] for predicting the profile of the surface and surface roughness values (Ra) as a function of feed (f), cutting tool geometry and tool errors. The research was focused on cutting with inserts that had a relatively large nose radius (r), and the influence of such tool errors were described in-depth as radial ($\varepsilon_r$) and axial ($\varepsilon_a$) runouts.

The investigations in many directions also point to the fact that face milling is characterized by inhomogeneity of roughness and its values differ in the different planes and surface elements of the surface. There are different roughness values measured in the direction of feed in the direction of the rotational axis of the tool and at different distances to the symmetry line. Theoretically, the highest roughness of the milled surface is in the symmetry plane. Therefore, in this study, the effect of increasing the feed is examined for the topography of the surface in the symmetry plane while analysing the change in the roughness values for single and multi-point milling operations. In the meantime, the axial setting errors of the inserts were also considered. For estimating roughness parameters, surface topography modelling is used that allows the determination and analysis of the theoretical values of 2D and 3D roughness parameters and that can take setting errors into account [20]. This article focuses on how the two- and three-dimensional surface roughness characteristics are affected when using a constant depth of cut. The considered 2D roughness parameters are the following: Ra—arithmetical average of the profile heights; Rz—average value of the absolute values of the heights of five highest profile peaks and the depths of five deepest valleys; Rq—root mean square average of the profile heights. The investigated 3D roughness parameters are the following: Sa—arithmetic mean height of the surface; Sz—maximum height of the surface; Sq—root mean squared height of the surface.

## 2. Materials and Methods

The aim of the conducted experiments was to investigate the effect of feed increase on the topography of the surface in face milling with one or more inserts by analysing the theoretical and real values of the 2D and 3D roughness parameters. The analysis is performed in the direction of feed in the rotational axis of the tool. The studies also take into account the axial run-outs of the inserts.

### 2.1. Investigation Method

The analysis of topography was performed by analysing the theoretical and real values of 2D and 3D roughness parameters and studying the relationship between the two value sets (Figure 1). The 2D and 3D surface roughness was measured on an AltiSurf 520 surface roughness tester device using a CL2 confocal chromatic probe. The data was evaluated using AltiMap software.

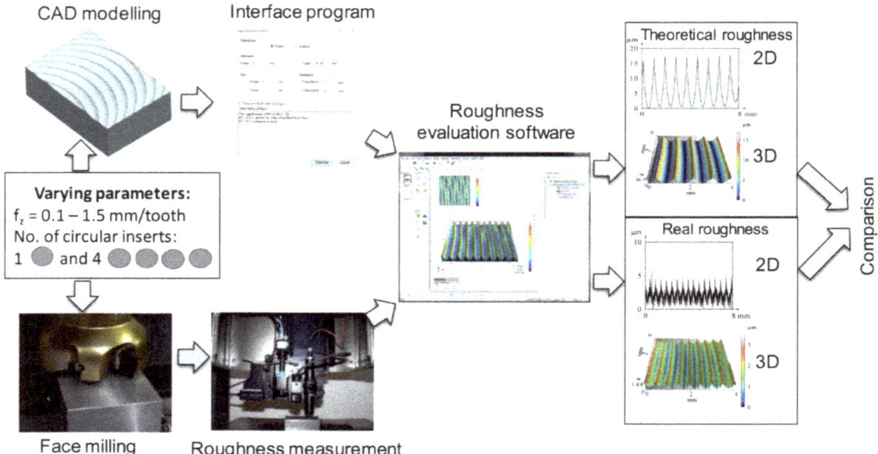



**Figure 1.** The applied investigation method.

## 2.2. Experimental Conditions

The milling tests were performed on C45 (1.0503 or AISI-1045) normalized plain carbon steel workpieces. The chemical composition of this material grade is summarized in Table 1. According to the material certification data sheet obtained from the manufacturer, the hardness of the material is 250 HB, its tensile strength is 700 Mpa, while its 0.2% proof stress ($Rp_{0.2}$) is 400 N/mm$^2$. The specimens were prepared as simple blocks with the dimensions of $100 \times 50 \times 50$ mm$^3$. The cutting data of the experiments and the characteristics of the machine tool and the tool can be found in Table 2.

**Table 1.** Chemical composition of the C45 material.

| C (%) | Si (%) | Mn (%) | P (%) | S (%) | Cr (%) | Ni (%) | Mo (%) |
|---|---|---|---|---|---|---|---|
| 0.430 | 0.170 | 0.740 | 0.009 | 0.032 | 0.070 | 0.080 | 0.013 |

**Table 2.** The applied technological data.

| Characteristic | One Insert | Four Inserts |
|---|---|---|
| $f_z$ (mm/tooth) | 0.5; 0.7; 0.9; 1.1; 1.3; 1.5 | 0.1; 0.3; 0.5; 0.7; 0.9 |
| n (1/min) | 950 | |
| $v_c$ (m/min) | 203 | |
| $a_p$ (mm) | 0.9 | |
| $a_e$ (mm) | 50 mm | |
| Milling head | Sandvik Coromant Coromill R200 milling head (R200-068Q27-12L) | |
| Milling insert | Circular insert, RCKT1204M0-PM 4230 iC = 12 mm | |
| Specimens | C45, $50 \times 50 \times 100$ mm$^3$ | |
| Machine | Perfect Jet MCV-M8 CNC vertical machining centre | |

During the experiments, a single insert was used in the first stage, while in the second series four inserts were clamped into the milling head. The position of inserts (cutting edges) was checked by a Zoller Hyperion tool pre-setter device, which has a measurement accuracy of around 2 μm and display accuracy of 1 μm. As the effect of radial run-outs on the machined surface is negligible with the investigated insert geometry, they were not investigated.

For the four inserts, the axial run-out values were the following: Insert1: 0 (the deepest one), Insert2: 14 μm, Insert3: 13 μm, Insert4: 4 μm.

## 2.3. Determination of Roughness Values by CAD Modelling

The analyses were carried out using a method for determining the theoretical values of surface roughness measurements [21]. Its essence is the CAD modelling of the machined surface. In this process, first the geometric model of the workpiece and the tool was created and then the imprints of the tool were made on the workpiece surface according to the kinematics characteristic of the process. The points of the surface (x, y and z coordinates)—i.e., the theoretical surface—were transferred to a professional surface topography analysis software using an interface software that performs evaluations based on standard two- or three-dimensional roughness parameters to the theoretical values. The AltiMap commercial surface roughness evaluation software was used to evaluate the desired two- and three-dimensional surface roughness parameters as well as visualization of roughness profiles.

One of the great advantages of this method is that both two-dimensional roughness profiles and three-dimensional surfaces can be generated in any measuring position and direction. The novelty of the method is that it is possible to analyse the complex surface topography created by any tool and the workpiece motion combination using the applied principle and the adjustment errors can be taken into account when using multi-point tools. Another great advantage is that it is relatively easy to introduce additional factors into the model. In this research, the option of modelling axial and radial differences between individual inserts has been used.

## 2.4. Examination of Roughness of the Machined (Milled) Surface

Examination of the machined surface is based on the 2D and 3D images from the topography and on the values of the measurements. For the evaluation of the surfaces of milling experiments the same topography measurement system is used; therefore, the data obtained by theoretical modelling can be validated on the same basis with the measured roughness profiles and values. The two-dimensional profiles, which are recorded in the feed direction at the centreline of the milling head, are measured in accordance with ISO 4287: 1997 [22] and ISO 4288: 1998 [23] standards. Three-dimensional surfaces are measured and evaluated in accordance with ISO 25178-2: 2012 [24].

## 3. Results

The tests were performed with either one or four circular inserts and the roughness parameters were analysed in 2D and 3D systems. Parameters measured in the 2D system: Ra, Rq and Rz. In the 3D system, their equivalents were measured: Sa, Sq and S10z. In each case a profile chart was taken in the 2D system and a topographic graph in the 3D system.

## 3.1. Roughness of the Modelled (Theoretical) Surface

Input parameters of the modelling (material quality, geometrical and technological parameters, etc.) were matched to the intended machining experiments and the ranges to be examined. After the modelling steps were completed, both the 2D and 3D profiles and the previously calculated theoretical values were available. From the theoretical values obtained by modelling, Figure 2 shows the results of three feed values for the simulation performed with a single circular insert (iC = 12 mm). Based on the presented topographic graphs, it can be seen, that profiles and three-dimensional surfaces obtained by modelling show a regular periodicity and accurately characterize the increase in surface roughness by increasing the feed.

A similar conclusion can be drawn from Figure 3, where four inserts are applied, with the exception that both profiles are more irregular. The periodicity can be observed here for one tool rotation. Within a revolution, traces of the inserts vary in depth, which is the consequence of the setting error (run-out). If there were no setting errors, then the profiles theoretically would be the same for the same feed per tooth when using one and four inserts. At this time and in this sense, cutting with a single insert can be considered as machining without a setting error.

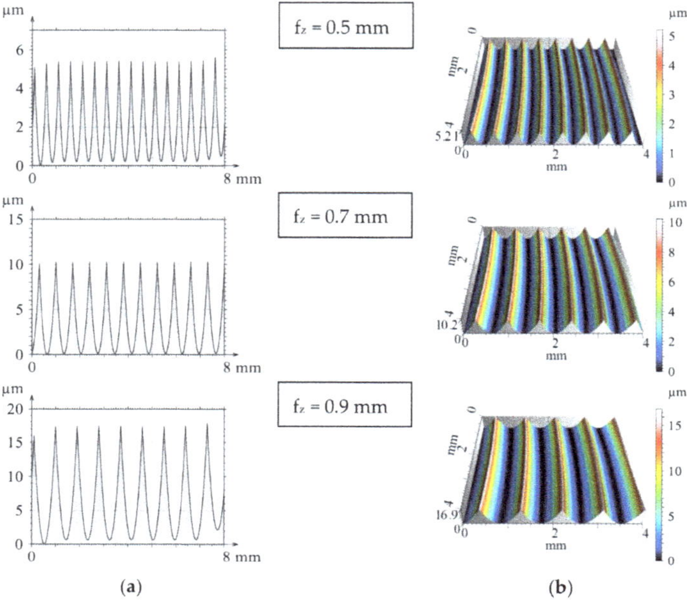

**Figure 2.** The modelled theoretical 2D (**a**) and 3D (**b**) surfaces when applying a single insert.

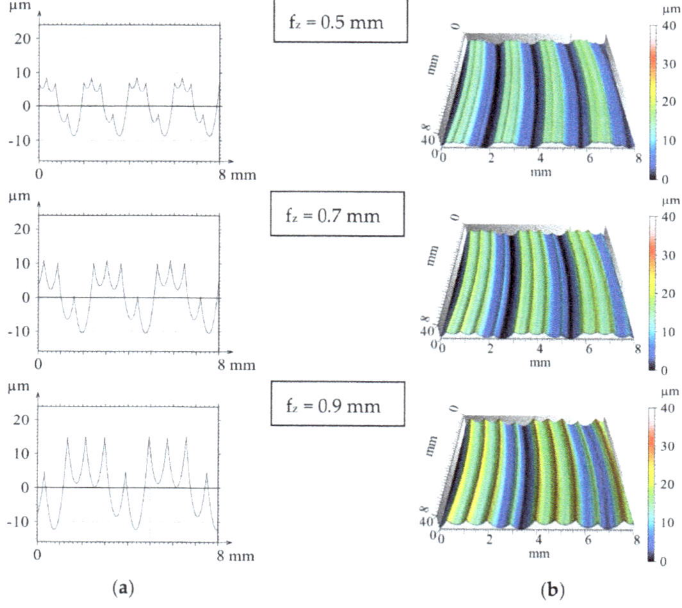

**Figure 3.** Modelled theoretical 2D (**a**) and 3D (**b**) surfaces when using four inserts.

## 3.2. Roughness of Milled Surfaces

Figures 4 and 5 show the measured roughness of the surfaces machined at the same feed but with one and four inserts respectively. If there were no insert run-outs, then the theoretical roughness would be the same in these cases. However, there are no multi-point milling tools without runout in the reality, so these should be considered in the modelling phase as well.

By comparing the theoretical and the measured profiles, it can be stated that the profiles are in good agreement and with the increasing value of the feed the theoretical profile more closely follows the real profile. This is typical for both single- and four-insert milling. In the case of four circular inserts, imprints of the successive non-coplanar inserts are clearly visible.

From the theoretical and measured values of roughness parameters, the values of Ra, Rq, Rz and Sa, Sq, S are summarized in Tables 3 and 4. In the tables bold values are used to indicate the same feed values for both cases.

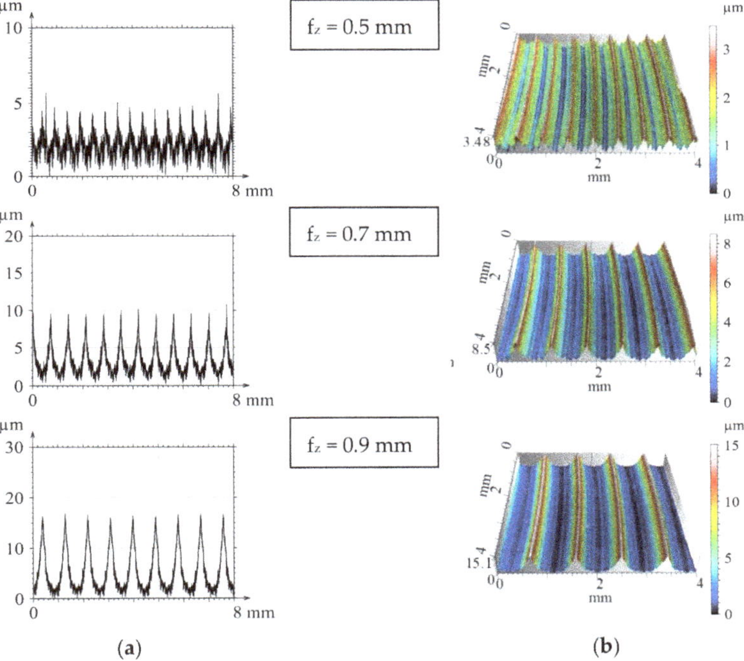


**Figure 4.** Roughness profiles (**a**) and topographic images (**b**) of face milling with one insert.

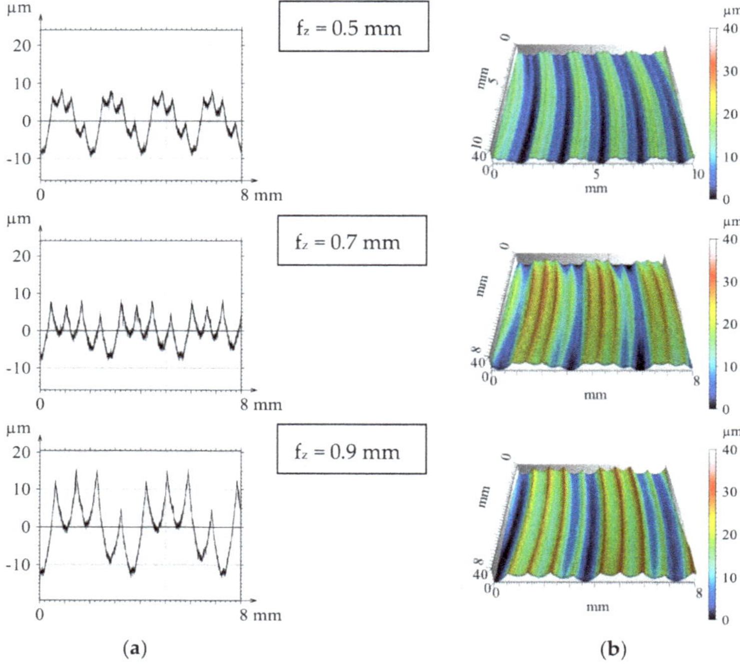

**Figure 5.** Milled 2D profiles (**a**) and 3D surfaces (**b**) when four inserts were applied.

**Table 3.** Roughness values on surfaces milled by a single insert.

| $f_z$ (mm) | Theoretical Roughness 1 Insert | | | | | | Measured Data 1 Insert | | | | | |
|---|---|---|---|---|---|---|---|---|---|---|---|---|
| | Ra μm | Sa μm | Rq μm | Sq μm | Rz μm | Sz μm | Ra μm | Sa μm | Rq μm | Sq μm | Rz μm | Sz μm |
| 0.5 | 1.34 | 1.33 | 1.56 | 1.55 | 5.2 | 5.21 | 0.62 | 0.62 | 0.79 | 0.785 | 5.21 | 3.48 |
| 0.7 | 2.64 | 2.63 | 3.07 | 3.05 | 10.2 | 10.2 | 1.56 | 1.56 | 1.97 | 1.96 | 9.94 | 8.5 |
| 0.9 | 4.33 | 4.26 | 5.01 | 4.96 | 16.8 | 16.9 | 3.5 | 3.35 | 4.21 | 4.12 | 16.7 | 15.1 |
| 1.1 | 6.27 | 6.62 | 7.29 | 7.62 | 24.6 | 25.1 | 5.65 | 5.61 | 6.49 | 6.51 | 23.9 | 22.3 |
| 1.3 | 8.29 | 9.03 | 9.69 | 10.5 | 33 | 35.2 | 7.91 | 8.36 | 9.21 | 9.69 | 32.9 | 34.5 |
| 1.5 | 12 | 12.4 | 14 | 14.3 | 47.1 | 47.1 | 11.7 | 12.2 | 13.7 | 14.1 | 48.9 | 48 |

**Table 4.** Roughness values on surfaces milled by four inserts.

| $f_z$ (mm) | Theoretical Roughness 4 Inserts | | | | | | Measured Data 4 Inserts | | | | | |
|---|---|---|---|---|---|---|---|---|---|---|---|---|
| | Ra μm | Sa μm | Rq μm | Sq μm | Rz μm | Sz μm | Ra μm | Sa μm | Rq μm | Sq μm | Rz μm | Sz μm |
| 0.1 | 0.856 | 0.851 | 0.994 | 0.989 | 3.33 | 3.3 | 1.05 | 1.14 | 1.35 | 1.41 | 6.41 | 6.99 |
| 0.3 | 4.04 | 4.26 | 4.62 | 4.85 | 14 | 14.7 | 3.66 | 3.79 | 4.26 | 4.42 | 16.5 | 16.4 |
| 0.5 | 5.11 | 5.04 | 5.52 | 5.47 | 17.2 | 18 | 4.57 | 4.78 | 5.09 | 5.3 | 19.2 | 19.1 |
| 0.7 | 5.47 | 5.32 | 6.13 | 6 | 21.5 | 21.4 | 5 | 5.75 | 5.74 | 6.81 | 23.3 | 29.5 |
| 0.9 | 5.97 | 6.12 | 7.27 | 7.4 | 27.3 | 29 | 5.83 | 6.26 | 7.08 | 7.49 | 30.6 | 30.4 |

## 4. Discussion

The analysis should be started by comparing the roughness values of milled surfaces machined by one and four inserts. In both cases, values of all parameters increase with increasing feeds and the change is of the same nature. It was also observed, that the Ra and Rq parameters as well as the Sa and Sq ones are varying in the same way and their values are very close to each other. So, only one of them, namely the much more widely used Ra (and Sa in 3D) will be evaluated in the following. Along with the same feed per tooth, the roughness is always lower for cutting with a single insert. After machining with four inserts, the roughness of the surface has increased considerably compared to machining with a single insert. For $f_z = 0.5$ mm, the theoretical values are 3.81 times greater for Ra, 3.31 times for Rz, 3.79 times for Sa and Sz is 3.45 times larger when machining with four inserts. Table 5 shows that the differences in measured values are even greater.

**Table 5.** The degree of increase in roughness values when changing the number of inserts.

| $f_z$ (mm) | Theoretical | | | | Measured | | | |
|---|---|---|---|---|---|---|---|---|
| | Ra | Rz | Sa | Sz | Ra | Rz | Sa | Sz |
| 0.5 | 3.81 | 3.31 | 3.79 | 3.45 | 7.37 | 3.69 | 7.71 | 5.49 |
| 0.7 | 2.07 | 2.11 | 2.02 | 2.10 | 3.21 | 2.34 | 3.69 | 3.47 |
| 0.9 | 1.38 | 1.63 | 1.44 | 1.72 | 1.67 | 1.83 | 1.87 | 2.01 |

When using feed per tooth of $f_z = 0.9$ mm, this difference is reduced by nearly 1.5-fold and the difference between the measured and the theoretical values is smaller. This means that the accurate setting of the insert is at least as important as changing feed values. It also shows that due to smaller deviations at higher feed, roughness can be more easily predicted. It can also be stated that the greater the value of the feed per tooth, the better the correlation between theoretical and the real roughness. One explanation for this is that in the range of lower feed values the additional effects of chip removal (such as vibrations in the milling cutter, tearing of the workpiece material during the chip removal, built-up edge, defects in the homogeneity of the workpiece material, undeformed chip thickness and tool wear) and the influence of the edge radius and the roughness of the cutting edge are greater in proportion.

Analysing the complete investigation range of one and four inserts, it is also found that the difference between the theoretical and the real roughness values is smaller for one insert than when cutting with four inserts (Figures 6 and 7).

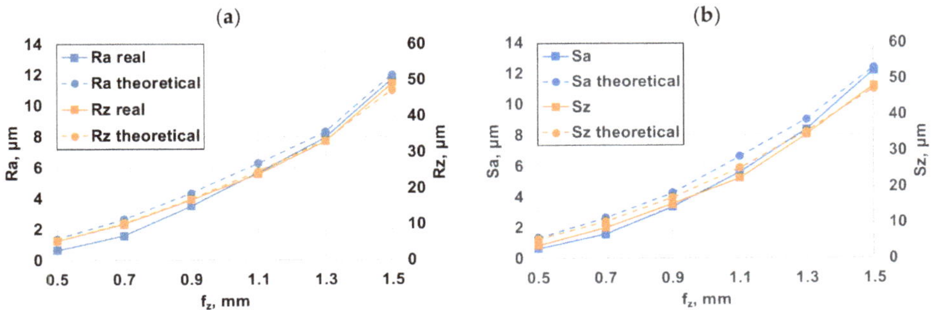


**Figure 6.** The roughness results of single-insert face milling: (**a**) 2D roughness parameters; (**b**) 3D roughness parameters.

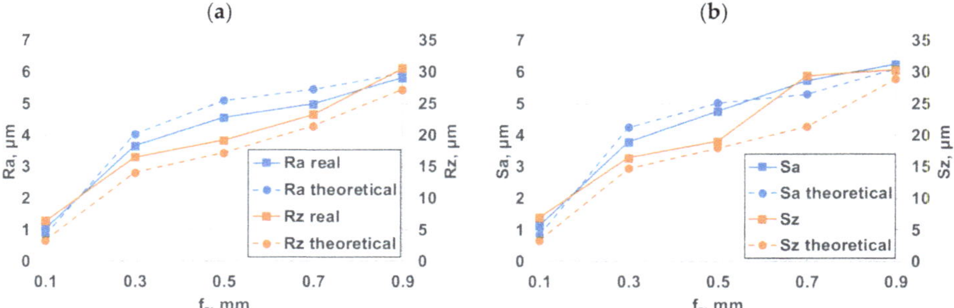

**Figure 7.** Comparison diagrams of four-insert roughness: (**a**) 2D roughness parameters; (**b**) 3D roughness parameters.

However, it is worth noting that, especially with respect to the Ra and Sa parameters, the theoretical methods described above have provided slightly greater roughness values than those found on the measured surfaces. This is caused by the relatively large radius (related to the roughness) and here the plastic deformation is more dominant because the cutting effect (material separation) on the tool edge is insufficiently applied, thus the roughness profile is distorted related to the theoretical one. Previous experience has shown that—as it was already mentioned before—the Ra and Rq parameters vary in the same way and their values are almost the same. Therefore, using only one of them—the more general Ra—is enough during the measurements. It has also been observed that neither Ra nor Rq reflects the change of roughness with sufficient sensitivity. Rz responds much more sensitively and more accurately to roughness than Ra. The tendency in the 2D profiles seems to increase the number of deep grooves forming by increasing the feed per tooth ($f_z$). This is unfavourable from a fatigue stress viewpoint but it is beneficial for oil storage. On the 3D profile pictures, however, there is a characteristic embossment which can be explained by the setting errors of the edges. In 3D images, the microgeometric shape of the surface can be said to be regular. A "bump" corresponds to a revolution of the tool.

When studying the data, it is apparent that the roughness parameters are considerably worse in the case of the four-insert milling tool with unchanged feed per tooth. This significant deterioration is clearly due to the run-out errors of the successive inserts. The result of axial run-out is that the inserts engage at different depths, resulting in a similar change in roughness amplitudes. By illustrating the measurement results, it can be established that when using four inserts, roughness can be increased up to seven times at low feed rates. This result is worse than expected. The main reason for this significant roughness deterioration is the setting inaccuracy of the inserts. The roughness peaks and the periodicity that correspond to a revolution of the tool can be easily tracked on the 3D topographic figures.

## 5. Summary and Outlook

The method of the study described in this article—simultaneously examining the roughness of the modelled surface and the machined surface—provides a good opportunity to analyse the topography of the machined surface. Based on the experiments carried out it can be stated that the accuracy of the approximation increases with the increase in feed and therefore the application of the method in the case of high-feed milling is particularly effective. Overall, it is therefore suitable for determining the theoretical values of the surface roughness and for estimating the expected roughness of the surfaces machined under the given conditions. In the case of single-insert face milling (fly-cutting), the roughness is gradually deteriorating with the increasing of the feed per tooth $f_z$. This change is most strongly reflected by the change in roughness parameters Rz and Sz. By comparing these roughness values and the four-insert experimental results, it was found that, depending on the feed rate—for the investigated settings—the surface had a 1.44–7.71 times worse roughness, which can be explained by the run-outs of the inserts. Commercial milling heads and commercial inserts were used in the experiments. The insert setting errors can be much higher in case of using non-standard cutter heads with special design, for example, in Reference [25]. Since tool assembly may result in a run-out that has a significant impact on the topography, it is advisable to check the run-out every time the inserts are replaced. Comparative analysis of modelled and measured roughness data showed that there is good compliance between the two values. Therefore, the roughness of the surfaces milled with various feeds can be well estimated in advance.

**Acknowledgments:** The authors greatly appreciate the support of the National Research, Development and Innovation Office—NKFIH (No. of Agreement: OTKA K 116876). The described article was carried out as part of the EFOP-3.6.1-16-00011 "Younger and Renewing University—Innovative Knowledge City—institutional development of the University of Miskolc aiming at intelligent specialisation" project implemented in the framework of the Szechenyi 2020 program. The realization of this project is supported by the European Union, co-financed by the European Social Fund.

**Author Contributions:** János Kundrák and Csaba Felhő conceived and designed the experiments; Csaba Felhő performed the experiments; János Kundrák and Csaba Felhő analysed the data; Csaba Felhő contributed reagents/materials/analysis tools; János Kundrák wrote the paper.

**Conflicts of Interest:** The authors declare no conflict of interest.

## References


1. He, C.L.; Zong, W.J.; Zhang, J.J. Influencing factors and theoretical modelling methods of surface roughness in turning process: State-of-the-art. *Int. J. Mach. Tools Manuf.* **2018**. [CrossRef]
2. Tomov, M.; Kuzinovski, M.; Cichosz, P. Modelling and prediction of surface roughness profile in longitudinal turning. *J. Manuf. Process.* **2016**, *24*, 231–255. [CrossRef]
3. Grzenda, M.; Bustillo, A. The evolutionary development of roughness prediction models. *Appl. Soft Comput.* **2013**, *13*, 2913–2922. [CrossRef]
4. Colak, O.; Kurbanoglu, C.; Kayacan, M.C. Milling surface roughness prediction using evolutionary programming methods. *Mater. Des.* **2007**, *28*, 657–666. [CrossRef]
5. El-Sonbaty, I.A.; Khashaba, U.A.; Selmy, A.I.; Ali, A.I. Prediction of surface roughness profiles for milled surfaces using an artificial neural network and fractal geometry approach. *J. Mater. Process. Technol.* **2008**, *200*, 271–278. [CrossRef]
6. Tseng, T.-L.; Konada, U.; Kwon, Y. A novel approach to predict surface roughness in machining operations using fuzzy set theory. *J. Comput. Des. Eng.* **2016**, *3*, 1–13. [CrossRef]
7. Rawangwong, S.; Chatthong, J.; Boonchouytan, W.; Burapa, R. Influence of Cutting Parameters in Face Milling Semi-Solid AA 7075 Using Carbide Tool Affected the Surface Roughness and Tool Wear. *Energy Procedia* **2014**, *5*, 448–457. [CrossRef]
8. Zeroudi, N.; Fontaine, M. Prediction of machined surface geometry based on analytical modelling of ball-end milling. *Procedia CIRP* **2012**, *1*, 108–113. [CrossRef]
9. Miko, E.; Nowakowski, L. Analysis and Verification of Surface Roughness Constitution Model after Machining Process. *Procedia Eng.* **2012**, *39*, 395–404. [CrossRef]
10. Ozcelik, B.; Bayramoglu, M. The statistical modelling of surface roughness in high-speed flat end milling. *Int. J. Mach. Tools Manuf.* **2006**, *46*, 1395–1402. [CrossRef]
11. Karkalos, N.E.; Galanis, N.I.; Markopoulos, A.P. Surface roughness prediction for the milling of Ti-6Al-4V ELI alloy with the use of statistical and soft computing techniques. *Measurement* **2016**, *90*, 25–35. [CrossRef]
12. Mikó, B.; Beno, J.; Mankova, I. Experimental Verification of Cusp Heights when 3D Milling Rounded Surfaces. *Acta Polytech. Hung.* **2012**, *9*, 101–116.
13. Shyha, I.; Gariani, S.; El-Sayed, M.A.; Huo, D. Analysis of Microstructure and Chip Formation When Machining Ti-6Al-4V. *Metals* **2018**, *8*, 185. [CrossRef]
14. Klickap, E.; Yardimeden, A.; Celik, Y.H. Mathematical Modelling and Optimization of Cutting Force, Tool Wear and Surface Roughness by Using Artificial Neural Network and Response Surface Methodology in Milling of Ti-6242S. *Appl. Sci.* **2017**, *7*, 1064. [CrossRef]

15. Selaimia, A.-A.; Yallese, M.A.; Bensouilah, H.; Meddour, I.; Khattabi, R.; Mabrouki, T. Modelling and optimization in dry face milling of X2CrNi18-9 austenitic stainless steel using RMS and desirability approach. *Measurement* **2017**, *107*, 53–67. [CrossRef]
16. Nguyen, H.-T.; Hsu, Q.-C. Surface Roughness Analysis in the Hard Milling of JIS SKD61 Alloy Steel. *Appl. Sci.* **2016**, *6*, 172. [CrossRef]
17. Baek, D.K.; Ko, T.J.; Kim, H.S. Optimization of feedrate in a face milling operation using a surface roughness model. *Int. J. Mach. Tools Manuf.* **2001**, *41*, 451–462. [CrossRef]
18. Schmitz, T.L.; Couey, J.; Marsh, E.; Mauntler, N.; Hughes, D. Runout effects in milling: Surface finish, surface location error and stability. *Int. J. Mach. Tools Manuf.* **2007**, *47*, 841–851. [CrossRef]
19. Franco, P.; Estrems, M.; Faura, F. Influence of radial and axial runouts on surface roughness in face milling with round insert cutting tools. *Int. J. Mach. Tools Manuf.* **2004**, *44*, 1555–1565. [CrossRef]
20. Felho, C. Investigation of Surface Roughness in Machining by Single and Multi-Point Tools. Ph.D. Thesis, Otto von Guericke University, Magdeburg, Germany, 2014.
21. Felho, C.; Karpuschewski, B.; Kundrak, J. Surface roughness modelling in face milling. *Procedia CIRP* **2015**, *31*, 136–141. [CrossRef]
22. International Organization for Standardization (ISO). *Geometrical Product Specifications (GPS)—Surface Texture: Profile Method—Terms, Definitions and Surface Texture Parameters*; ISO 4287; ISO: Geneva, Switzerland, 1997.
23. International Organization for Standardization (ISO). *Geometrical Product Specifications (GPS)—Surface Texture: Profile Method—Rules and Procedures for the Assessment of Surface Texture*; ISO 4288; ISO: Geneva, Switzerland, 1998.
24. International Organization for Standardization (ISO). *Geometrical Product Specifications (GPS)—Surface Texture: Areal—Part 2: Terms, Definitions and Surface Texture Parameters*; ISO 25178-2; ISO: Geneva, Switzerland, 2012.
25. Beno, J.; Mankova, I.; Vrabel, M.; Karpuschewski, B.; Emmer, T.; Schmidt, K. Operation Safety and Performance of Milling Cutters with Shank Style Holders of Tool Inserts. *Procedia Eng.* **2012**, *48*, 15–23. [CrossRef]

*Article*

# Comparative Analysis of Machining Procedures

Janos Kundrak [1], Viktor Molnar [2,*] and Istvan Deszpoth [1]

[1]  Institute of Manufacturing Science, University of Miskolc, Miskolc-Egyetemvaros, 3515 Miskolc, Hungary;
    janos.kundrak@uni-miskolc.hu (J.K.); istvan.deszpoth@uni-miskolc.hu (I.D.)
[2]  Institute of Management Science, University of Miskolc, Miskolc-Egyetemvaros, 3515 Miskolc, Hungary
*   Correspondence: szvmv@uni-miskolc.hu

Received: 28 February 2018; Accepted: 22 March 2018; Published: 28 March 2018

**Abstract:** The in-depth analysis of cutting procedure is a topic of particular interest in manufacturing efficiency because in large-scale production the effective use of production capacities and the revenue-increasing capacity of production are key conditions of competitiveness. That is why the analysis of time and material removal rate, which are in close relation to production, are important in planning a machining procedure. In the paper three procedures applied in hard cutting are compared on the basis of these parameters and a new parameter, the practical parameter of material removal rate, is introduced. It measures not only the efficiency of cutting but also that of the whole machining process because it includes the values measured by time analysis as well. In the investigations the material removal rate was analyzed, first on the basis of geometrical data of the component. After that different machining procedures (hard machining) were compared for some typical surfaces. The results can give some useful indications about machining procedure selection.

**Keywords:** hard turning; grinding; combined procedure; material removal rate; operation time

---

## 1. Introduction

Machining procedures for machining industry components have developed rapidly in the last decade thanks to new, powerful machines with high load bearing capacities, the materials used in their structures, and their ever more advanced control systems. This technological development facilitates the machining of components with higher accuracy and better surface quality in shorter time. Due to the shorter machining time, more components can be produced within a given time and therefore more profit can be earned. Exact determination of time parameters directly connected to machining is critical, considering that the time decrease of one component is significant if the sum of these time values is considered over a year. This can result in a high revenue surplus in mass production. In the time of Industry 4.0, when the conception of computer controlled automated plants is being extended, automatic data collection and analysis can be utilized to help in intelligent decision-making. This means the time data of production processes are available in a shorter time and at a higher level of accuracy [1–4]. For a given technology the machining time and cost can basically be optimized by the cutting data (cutting speed, depth-of-cut, feed, etc.). Additionally, rationalization of supporting activities of a production process can decrease the times connected directly or indirectly to the machining. This means that decreases can be made in the preparation time or the time needed for material handling among the workplaces, for instance. Several plant management solutions for this purpose have evolved in recent decades, e.g., lean production, six sigma or the theory of constraints [5,6].

Beyond changes in cutting data and the rationalization of manufacturing organization, another essential factor in increasing of production efficiency is the choice of procedures or procedure versions used in machining a given component. If the same surface quality and accuracy can be achieved by two procedures using completely different technologies, the two versions [7] can be considered as perfectly replaceable alternatives from the point of view of machining [8]. In addition to that, of course, the investment volume and other costs related to the machine and equipment needs of the procedures have to be compared, like a skilled workforce or maintenance. Three machining procedures are compared in the paper for finishing hardened surfaces. Conventional grinding is considered as the base point and hard turning and a combined procedure are compared to it. Hard turning and the combined procedure are new solutions in machining but several aspects of conventional grinding are still researched [9,10] In the latter procedure hard turning and grinding are applied in one clamping of the workpiece in order to exploit the advantages of both procedures. This begins with hard turning, whose material removal efficiency is relatively high but which forms a periodic topography, which is not always suitable for requirements for in-built components. Thus, grinding is necessary after cutting with a single-point cutting tool. In this case including grinding in the combined procedure reduces the material removal efficiency compared to hard turning by only a small extent, because to remove periodic topography it is sufficient to remove only a minimal depth of material ($R_{max}$ scale of the hard turned surface). As this is done in one clamping with hard turning, it leads to little increase in machining time, or even a decrease [11,12].

## 2. Method of Analysis

The investigation was carried out for gear wheel finish machining operations. These machine parts have three geometrically distinct surfaces that must be machined. Thus, with the introduced calculation method, the analysis can be applied reliably and simply not only for the different procedures but for the different typical surfaces too. Our earlier experiments dealt with the increase in material removal efficiency achievable by changing technological data, comparing also these three procedures [13,14]. In this study the machining time ($T_m$), the operation time ($T_{op}$), the theoretical value of material removal rate ($Q_w$) and the practical material removal rate ($Q_{wp}$) were analyzed in the three machining procedures. After that the effect of changes in geometrical data (bore length and diameter) on the practical material removal rate in machining internal cylindrical surface were analyzed. In the first step the efficiency of material removal was analyzed using the theoretical material removal rate. This parameter can be determined on the basis of the cutting and geometrical data of a component, but a more accurate picture of efficiency can be gained if a time base characterized by real production circumstances is considered.

For this purpose we define a parameter that considers these times, naming it of the practical material removal rate. This practical indicator is also suitable for characterizing production processes (machining and connecting production organization). Thus, calculations by this parameter are suitable to support more complex technological decisions. The logic of this is outlined in Figure 1.

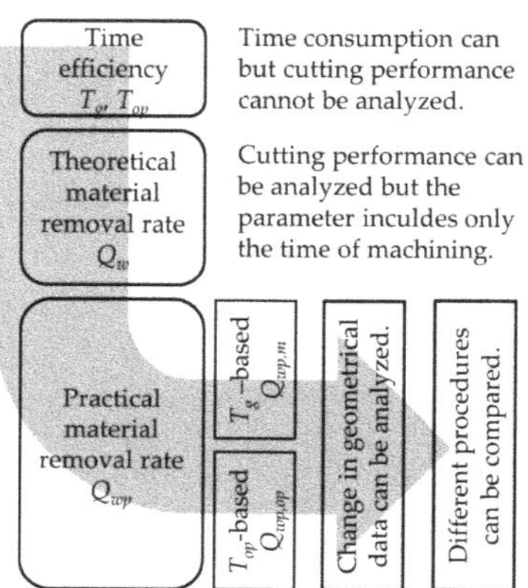


**Figure 1.** Logic of the analysis.

## 3. Material Removal Efficiency Measurement Parameters

### 3.1. Time Parameters

A relatively inflexible feature of machining is the machining time, i.e. the time the workpiece spends in machining on a machine tool. Reduction of this is possible as long as the quality requirements can still be fulfilled. The other useful parameter is the operation time, which includes the preparation and finishing time of machining, the supplementary times, and other times of operations and processes that are directly needed to produce a component. Projecting the times beyond machining time to one workpiece can also be significant, therefore the reduction of such times is also important for the planners of production process. In our experiments we analyzed machining of the internal cylindrical surface (ICS), plain surface (PS) and shaped surface (SS)—a cone. In the analysis hard turning performed by a single-point tool (ICS, PS, SS), face grinding (PS) and in-feed grinding (ICS, SS) were compared. Calculations were performed by the following formulas. The variables of the formulas are summarized in Table 1.

Bore grinding (roughing and smoothing passes):

$$T_m = \frac{2L}{v_{fL,R}} \cdot \frac{Z_R}{a_{e,R}} + \frac{2L}{v_{fL,S}} \cdot \left( \frac{Z_S}{a_{e,S}} + i_{so} \right) \tag{1}$$

In-feed bore and cone grinding (roughing and smoothing passes):

$$T_m = \frac{Z_A}{v_{fR,A}} + \frac{Z_R}{v_{fR,R}} + \frac{Z_S}{v_{fR,S}} + t_{so} \tag{2}$$

Face grinding (roughing and smoothing passes):

$$T_m = \frac{1}{n_w} \cdot \left( \frac{Z_R + 0.1}{a_{p,R}} + \frac{Z_S}{a_{p,S}} + i_{so} \right) \tag{3}$$

Hard turning of bore, cone and face (roughing and smoothing passes):

$$T_m = \frac{L\prime}{f_R n_w} + \frac{L\prime}{f_S n_w} = \frac{d_w L\prime \pi}{1000 v_c f_R} + \frac{d_w L\prime \pi}{1000 v_c f_S} \tag{4}$$

Combined procedure (hard turning and then in-feed grinding):

$$T_m = \frac{d_w L\prime \pi}{1000 v_c f_R} + \frac{Z_A}{v_{fR,A}} + \frac{Z_S}{v_{fR,S}} + t_{so} \tag{5}$$

Several methods exist for the calculation of operation time of machining. Here we present the formulas applied in the plant where the analyzed gears are machined.

$$T_{op} = \frac{T_{prep}}{n} + T_{piece} \tag{6}$$

$$T_{piece} = T_{base} + T_{suppl} \tag{7}$$

$$T_{base} = T_m + T_{manip} \tag{8}$$

$$T_{suppl} = k T_{base} \tag{9}$$

$$T_{op} = \frac{T_{prep}}{n} + (1 + k)(T_m + T_{manip}) \tag{10}$$

where $T_{op}$ is operation time, $T_{prep}$ time of preparation and finish, $T_{piece}$ piece time, $T_{base}$ base time, $T_{suppl}$ supplementary time, $T_m$ machining time, $T_{manip}$ workpiece manipulation time, and $k$ is a coefficient whose value here is 0.2.

**Table 1.** Geometrical and cutting data applied in the formulas of machining time.

| Symbol | Description |
| --- | --- |
| $L$ | bore length |
| $L'$ | bore length + tool overrun |
| $Z_R; Z_S; Z_A$ | roughing, smoothing and air grinding allowance |
| $v_{fL,R}; v_{fL,S}$ | traverse feed rate (roughing, smoothing) |
| $a_{e,R}; a_{e;S}$ | depth-of-cut in grinding (roughing, smoothing) |
| $i_{so}$ | number of sparking out revolutions |
| $v_{fR,R}; v_{fR,S}; v_{fR,A}$ | radial feed rate in in-feed grinding (roughing, smoothing, air grinding) |
| $n_w$ | revolution per minute of the workpiece |
| $a_{p,R}; a_{p,S}$ | depth-of-cut (roughing, smoothing) |
| $f_R, f_S$ | roughing and smoothing feed |
| $d_w$ | workpiece diameter |
| $v_c$ | cutting speed in turning |
| $t_{so}$ | time of sparking out |

## 3.2. Material Removal Rate

The theoretical material removal rate ($Q_w$) defines what material volume can be removed from the surface in a time unit. It does not consider the time of machining during which the workpiece is not physically cut (e.g., manipulation). That is why it only shows the effect of change of technological data. This parameter can be calculated in the different procedures. Since it does not include all factors (e.g., sparking out, tool overrun), it cannot be considered as a sufficiently exact parameter in comparing different procedures. The calculation method of the theoretical parameter is summarized in Table 2.

**Table 2.** Calculation of the theoretical parameter of the material removal rate in different procedures.

| Procedure | $Q_w$ [mm$^3$/s] |
| --- | --- |
| Face grinding | $a_p b v_w$ |
| Bore grinding | $a_e f v_w$ |
| In-feed bore grinding<br>In-feed cone grinding | $a_p d_w \pi v_{fR}$ |
| Face turning<br>Boring<br>Cone turning | $a_p f v_c$ |

In the calculation of the practical parameter the removed material volume is divided by a certain time data ($T_x$) characterizing the production of a surface element/surface/component. This value will be the specific material volume, i.e. the parameter measures the material removal rate while considering the time components of machining. If the preferred technological decision factor is the specific material volume removed while the workpiece is clamped, the machining time is considered. If other supplementary times significantly influence the production, the piece time is considered, and so on. In our analyses operation time was included in the calculations. Since the operation time is what expresses the real time consumption, we can also draw conclusions on the efficiency of the whole production process. The practical material removal rates calculated by the operation time are given by Equations (11)–(13).

Bore:

$$Q_{wp,op} = \frac{Ld\pi Z}{60T_x} \tag{11}$$

Face:

$$Q_{wp,op} = \frac{L(d-L)\pi Z}{60T_x} \tag{12}$$

Cone:

$$Q_{wp,op} = \frac{L\cos\alpha(d - L\,tg\alpha)\pi Z}{2\cdot 60T_x} \tag{13}$$

where $L$ is machined bore length, $d$ workpiece diameter, $Z$ allowance, $\alpha$ half cone-angle, and $T_x$ considered time.

The rate of theoretical to practical parameters shows the rate of extra time necessary for machining compared to the machining time. We note that the surface rate is a similar parameter. That differs from the material removal rate in showing the specific area of removed surface. The value of the parameter is equal to that of the material removal rate if material removal is performed in one pass.

## 4. Basic Data of Comparison Analyses

In the study calculations were made for the machining of an analyzed gear wheel. The component is comprised of one plain surface, one conical surface and one internal cylindrical surface to be machined. The material of the component was 16MnCr5 (HRC 62). Its geometrical and cutting data are summarized in Figure 2 and Table 3, where the symbols are:

- Procedures: P1: grinding; P2: hard turning; P3: combined procedure (P3/1: hard turning, P3/2: in-feed grinding)
- Surfaces: S1: face; S2: bore; S3: cone

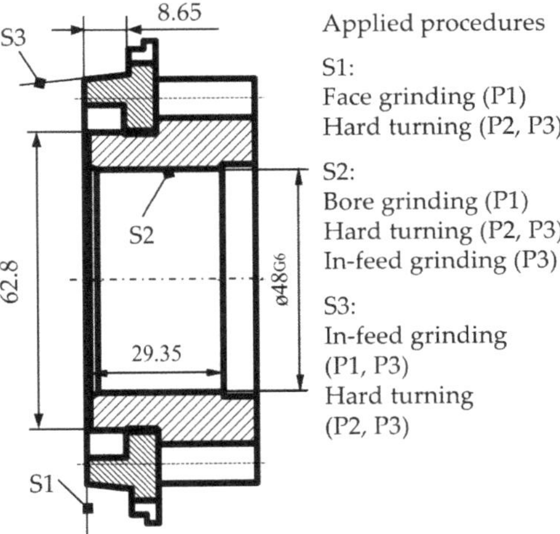


**Figure 2.** Geometrical and cutting data.

**Table 3.** Cutting data.

**Surface 1 (S1)—Face**

| P1 | |
|---|---|
| $a_{e,R}$: 0.03 mm | |
| $a_{e,S}$: 0.01 mm | |
| $v_w$: 15 m/min | |
| $v_c$: 30 m/s | |
| $i_{so}$: 8 | |
| **P2** | |
| $a_p$: 0.3 mm | |
| $f$: 0.08 mm/rev | |
| $v_c$: 228 m/min | |
| **P3** | |
| $a_p$: 0.3 mm | |
| $f$: 0.08 mm/r | |
| $v_c$: 228 m/min | |

**Surface 2 (S2)—Bore**

| P1 | |
|---|---|
| $n_{k,R}$: 40 ds/min | $a_{e,R}$: 0.01 mm/ds |
| $n_{k,S}$: 36 ds/min | $a_{e,S}$: 0.001 mm/ds |
| $f_R$: 24.44 mm/r | $v_{fL,R}$: 2200 mm/min |
| $f_S$: 22.22 mm/r | $v_{fL,S}$: 2000 mm/min |
| $v_w$: 13.6 m/min | $i_{so}$: 16 |
| $v_c$: 29 m/s | |
| **P2** | **P3/1** |
| $a_{p,R}$: 0.25 mm | $a_p$: 0.25 mm |
| $a_{p,S}$: 0.05 mm | $f$: 0.15 mm/r |
| $f_R$: 0.15 mm/r | $v_c$: 180 m/min |
| $f_S$: 0.08 mm/r | |
| $v_c$: 180 m/min | |
| **P3/2** | |
| $v_c$: 40 m/s | $Z_N$: 0.04 mm |
| $v_w$: 86 m/min | $Z_S$: 0.01 mm |
| $v_{fR,R}$: 0.005 mm/s | $Z_A$: 0.27 mm |
| $v_{fR,S}$: 0.0033 mm/s | $t_{so}$: 5 s |
| $v_{fR,A}$: 0.1 mm/s | |

**Surface 3 (S3)—Cone**

| P1 | |
|---|---|
| $v_c$: 32 m/s | $t_{so}$: 6 s |
| $v_w$: 98 m/min | $Z_N$: 0.04 mm |
| $v_{fR,R}$: 0.009 mm/s | $Z_S$: 0.01 mm |
| $v_{fR,S}$: 0.003 mm/s | |
| **P2** | **P3/1** |
| $a_p$: 0.3 mm | $a_p$: 0.3 mm |
| $f$: 0.12 mm/r | $f$: 0.12 mm/r |
| $v_c$: 224 m/min | $v_c$: 224 m/min |
| **P3/2** | |
| $v_c$: 32 m/s | $Z_N$: 0.035 mm |
| $v_w$: 98 m/min | $Z_S$: 0.015 mm |
| $v_{fR,R}$: 0.008 mm/s | $Z_A$: 0.27 mm |
| $v_{fR,S}$: 0.003 mm/s | $t_{so}$: 6 s |
| $v_{fR,A}$: 0.1 mm/s | |

## 5. Results and Discussion

### 5.1. Theoretical Material Removal Rate

The theoretical values of the material removal rate are summarized in Table 4. Machining of the typical surfaces of the component cannot be compared directly with the theoretical values but certain conclusions can be made. For example the bore (S2) can be machined more effectively by hard turning (P2) than by grinding (P1) because 6.75 > 3.32 and 0.72 > 0.3 in the same time. Concerning the machining of the face (S1) the two procedures cannot be compared because in grinding the material removal is carried out in two passes and in hard turning in only one pass.

**Table 4.** Values of the theoretical material removal rate ($Q_w$).

| | | | | | | | | |
|---|---|---|---|---|---|---|---|---|
| P1 | | | | | | | | |
| S1 | R | 6.66 | S2 | R | 3.32 | S3 | R | 6.69 |
| S1 | S | 2.22 | S2 | S | 0.3 | S3 | S | 2.23 |
| P2 | | | | | | | | |
| S1 | 5.47 | | S2 | R | 6.75 | S3 | 8.06 | |
| S1 | 5.47 | | S2 | S | 0.72 | S3 | 8.06 | |
| P3/1 | | | | | | | | |
| S1 | 5.47 | | S2 | 6.75 | | S3 | 8.06 | |
| P3/2 | | | | | | | | |
| - | | | S2 | R | 20.62 | S3 | R | 10.44 |
| - | | | S2 | S | 13.61 | S3 | S | 3.91 |

### 5.2. Practical Parameter of Material Removal

In Table 5 the machining time, the operation time and the practical values of material removal rate ($Q_{wp,op}$) are summarized for a given component. The rate of machining time to operation time was analyzed and is illustrated in Figure 3a. Operation time is considered as 100 percent. While in grinding the operation time is 1.56 times higher than the machining time, this value is 1.7 in hard turning and 2.57 in the combined procedure. The arcs representing the procedures also give the absolute time values. This figure highlights that there is a relatively large difference between the machining and the operation times and that is why it is worth focusing on the role of operation time in efficiency analyses. The values of the practical material removal parameter are given in Figure 3b. The values of grinding and the combined procedure are lower than those of hard turning for the analyzed component. In the figure it can be seen that despite this great difference the $Q_{wp,op}$ value of the combined procedure exceeds that of grinding. It is noted that in selection of the procedures not only the machining efficiency but also the costs of the procedures have to be calculated (e.g., machine tool investment).

**Table 5.** Machining time, operation time and $Q_{wp,op}$.

| | P1 | P2 | P3 |
|---|---|---|---|
| $T_m$ | 3.59 | 0.74 | 1.05 |
| $T_{op}$ | 5.55 | 1.26 | 2.70 |
| $Q_{wp,op}$ | 7.62 | 33.63 | 15.65 |

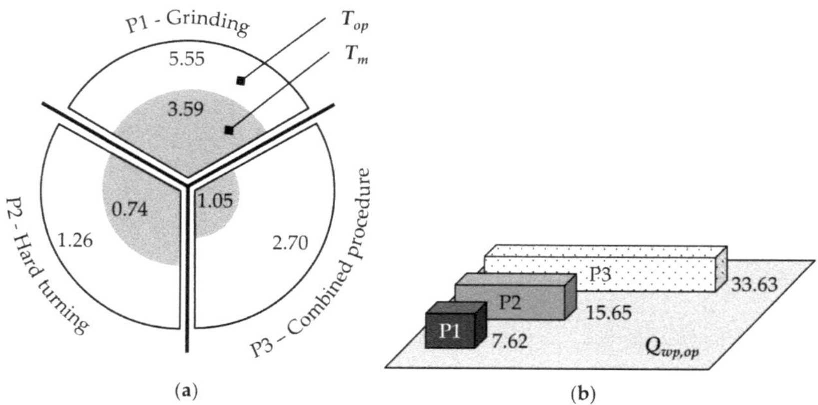



(a)        (b)

**Figure 3.** (**a**) rate of machining times within operation time ($T_{op}$ = 100%); (**b**) practical material removal rates of the three procedures.

## 5.3. Effect of Bore Length and Diameter

The practical material removal rates of the three procedures are summarized in Figures 4–6.

In Figure 7 the practical parameters of hard turning and the combined procedure are compared to those of grinding for different bore diameters and bore lengths when machining internal cylindrical surfaces, since this type of surfaces is hard to machine.

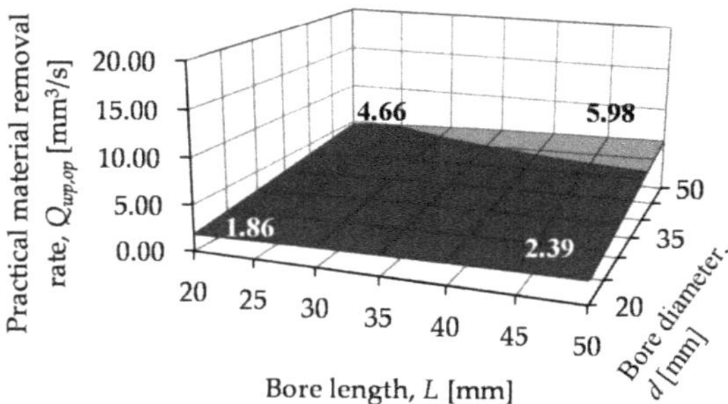



**Figure 4.** $Q_{wp,op}$ values of grinding.

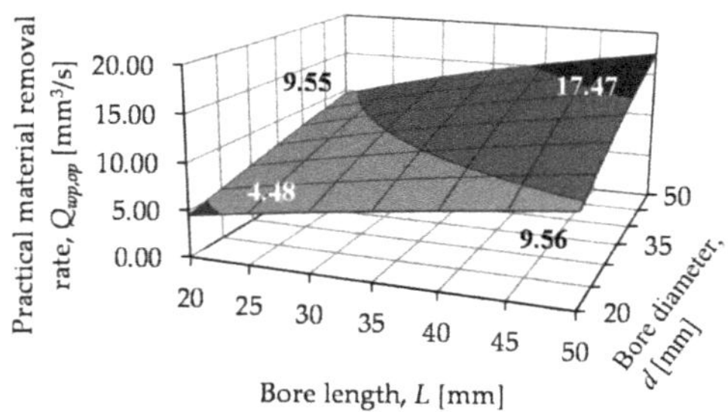



**Figure 5.** $Q_{wp,op}$ values of hard turning.

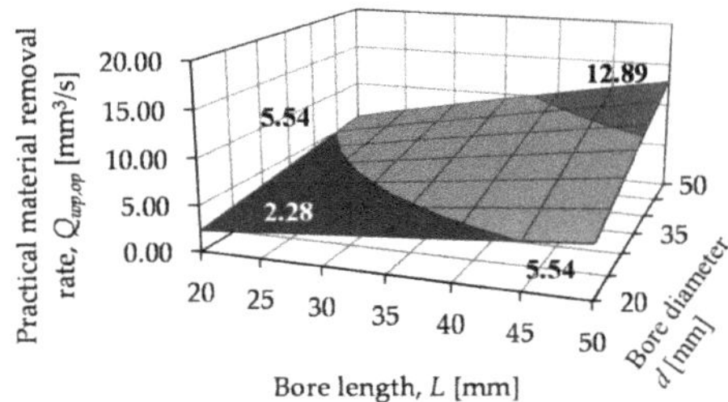

**Figure 6.** $Q_{wp,op}$ values of the combined procedure.

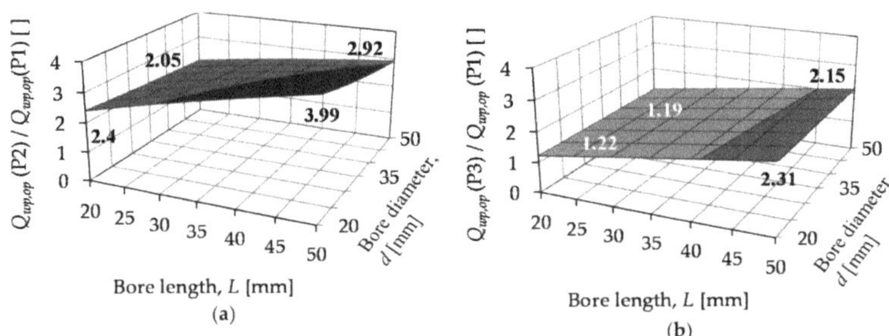

(a)

(b)

**Figure 7.** Rates of material removal rates of hard turning (**a**) and the combined procedure (**b**) compared to grinding.

In both procedures the difference is clear, namely the value of the practical material removal rate for hard turning exceeds that of the combined procedure but the value of the combined procedure is considerably better than that of grinding. Although the practical material removal rate is the highest in hard turning, the operation circumstances of the component can require the application of the combined procedure. In Figures 5 and 6 it can be seen that with the increase of both the diameter and the bore length the practical value of material removal rates increases. In hard turning when the bore length and the diameter are between 20 and 50 mm, the $Q_{wp,op}$ values are between 4.48 and 17.47 mm$^3$/s. In the combined procedure these values are between 2.33 and 13.12 mm$^3$/s. In Figure 7 the rates of $Q_{wp,op}$ compared to that of grinding (P1) are given for the introduced procedures (P2, P3). In hard turning the values of practical material removal rate are 2.05–3.99 times higher than those of the grinding. In the combined procedure these values are between 1.25 and 2.36. Application of the practical material removal rate analysis for geometrical data can supports the construction design of components.

## 6. Conclusions

Through analyzing the material removal rate the efficiency of machining procedures was calculated. This parameter characterizes the efficiency of the different machining procedures well but it does not facilitate the comparison of different procedures or procedure versions. To compare the procedures the practical parameter was analyzed. The $Q_{wp,op}$ value of grinding is 23 percent that of hard turning for the analyzed gear wheels. The practical material removal rate of the combined procedure is 47 percent that of hard turning. That is why the combined procedure is expedient to substitute for grinding if it is necessary to form a random topography. The practical value of the material removal rate $Q_{wp,op}$ is suitable for technological decision support within certain limits. It allows more analysis possibilities than the theoretical value because it provides more information than simple time data. However, it does not consider the direct and indirect costs related to machining, which can modify the decision made for the procedure choice. In the analysis of bore length and diameter, the increase of these two variables results in an increase in the practical material removal rate in case of fixed cutting data. In our tests both geometrical data were varied between 20 and 50 mm. In increasing both the bore length and bore diameter the $Q_{wp,op}$ increases 1.83–2.13-fold in hard turning. For smaller diameters and bore lengths the extent of increase is higher if only one parameter is changed at one time. In the combined procedure the $Q_{wp,op}$ value increases 2.32-2.42-fold when the diameter or the length are increased. This analysis can be useful in construction design of machined components. In summary, the method introduced here can be applied in the comparison of three separate aspects: different machining procedures; machining of components that contain different types of surfaces and identical types of surfaces with different geometrical values. These comparisons provide information not only about the efficiency of the applied procedures but also about the organization efficiency of production.

**Acknowledgments:** Project no. NKFI-125117 has been implemented with the support provided from the National Research, Development and Innovation Fund of Hungary, financed under the K_17 funding scheme.

**Author Contributions:** Janos Kundrak and Viktor Molnar conceived and designed the experiments; Viktor Molnar and Istvan Deszpoth performed the experiments; Istvan Deszpoth and Janos Kundrak analyzed the data; Janos Kundrak and Viktor Molnar wrote the paper.

**Conflicts of Interest:** The authors declare no conflict of interest.

## References


1. Hammer, M.; Somers, K.; Karre, H.; Ramsauer, C. Profit per hour as a target process control parameter for manufacturing systems enabled by Big Data analytics and Industry 4.0 infrastructure. *Procedia CIRP* **2017**, *63*, 715–720. [CrossRef]
2. Tamas, P.; Illes, B.; Dobos, P. Waste reduction possibilities for manufacturing systems in the industry 4.0. IOP Conference Series. *Mater. Sci. Eng.* **2016**, *161*, 1–8. [CrossRef]
3. Zhong, R.Y.; Xu, X.; Klotz, E.; Newman, S.T. Intelligent Manufacturing in the Context of Industry 4.0: A Review. *Engineering* **2017**, *3*, 616–630. [CrossRef]
4. König, W.; Berktold, A.; Koch, K.F. Turning versus grinding—A comparison of surface integrity aspects and attainable accuracies. *Ann. CIRP* **1993**, *42*, 39–43. [CrossRef]
5. Jacobs, B.W.; Swink, M.; Linderman, K. Performance effects of early and late Six Sigma adoptions. *J. Oper. Manag.* **2015**, *36*, 244–257. [CrossRef]
6. Castro, S.; Riedel, C. Assessment of the implementation of manufacturing excellence in a fiber based packaging manufacturing environment. *Procedia CIRP* **2017**, *63*, 113–118. [CrossRef]
7. Kundrak, J.; Mamalis, A.G.; Markopoulos, A. Finishing of hardened boreholes: Grinding or hard cutting? *Mater. Manuf. Process.* **2004**, *19*, 979–993. [CrossRef]
8. Klocke, F.; Brinkmeier, E.; Weinert, K. Capability profile of hard cutting and grinding process. *Ann. CIRP* **2005**, *54*, 22–54. [CrossRef]
9. Markopoulos, A.P.; Kundrak, J. FEM/AI models for the simulation of precision grinding. *Manuf. Technol.* **2016**, *16*, 384–390.
10. Dudas, L. Design of Nonconventional Grinding Wheels with Specialized CAD System. In Proceedings of the 4th IEEE International Symposium on Logistics and Industrial Informatics (LINDI 2012), Smolenice, Slovakia, 5–7 September 2012; pp. 21–26.
11. Tönshoff, H.K.; Arendt, C.; Ben Amor, R. Cutting of Hardened Steel. *Ann. CIRP* **2000**, *49*, 547–566. [CrossRef]
12. Waikar, R.A.; Guo, Y.B. A comprehensive characterization of 3D surface topography induced by hard turning versus grinding. *J. Mater. Process. Technol.* **2008**, *197*, 189–199. [CrossRef]
13. Kundrak, J. Alternative machining procedures of hardened steels. *Manuf. Technol.* **2011**, *11*, 32–39.
14. Kundrak, J.; Varga, G.; Deszpoth, I.; Molnar, V. Some aspects of machining of bore holes. *Appl. Mech. Mater.* **2013**, *309*, 126–132. [CrossRef]

 *machines*

*Article*

# Precision CNC Machining of Femoral Component of Knee Implant: A Case Study

**Angelos P. Markopoulos** *, **Nikolaos I. Galanis** , **Nikolaos E. Karkalos and Dimitrios E. Manolakos**

Section of Manufacturing Technology, School of Mechanical Engineering, National Technical University of Athens. Heroon Politechniou 9, 15780 Athens, Greece; ngalanis@central.ntua.gr (N.I.G); nkark@mail.ntua.gr (N.E.K.); manolako@central.ntua.gr (D.E.M.)
* Correspondence: amark@mail.ntua.gr; Tel.: +30-210-772-4299

Received: 13 February 2018; Accepted: 28 February 2018; Published: 2 March 2018

**Abstract:** The design and manufacturing of medical implants constitutes an active and highly important field of research, both from a medical and an engineering point of view. From an engineering aspect, the machining of implants is undoubtedly challenging due to the complex shape of the implants and the associated restrictive geometrical and dimensional requirements. Furthermore, it is crucial to ensure that the surface integrity of the implant is not severely affected, in order for the implant to be durable and wear resistant. In the present work, the methodology of designing and machining the femoral component of total knee replacement using a 3-axis Computer Numerical Control (CNC) machine is presented, and then, the results of the machining process, as well as the evaluation of implant surface quality are discussed in detail. At first, a preliminary design of the components of the knee implant is performed and the planning for the production of the femoral component is implemented in Computed Aided Manufacturing (CAM) software. Then, three femoral components are machined under different process conditions and the surface quality is evaluated in terms of surface roughness. Analysis of the results indicated the appropriate process conditions for each part of the implant surface and led to the determination of optimum machining strategy for the finishing stage.

**Keywords:** femoral component; medical implant; total knee replacement; CNC machining; implant machining; sculptured surface; bio-engineering; surface quality

---

## 1. Introduction

Partial or total replacement of a human knee with an implant can be achieved by a common surgical procedure, namely knee arthroplasty or knee replacement. Only in United States, over 500,000 knee replacement surgeries are reported each year [1], which are mostly for patients between 50 and 80 years old. One of the main reasons for the popularity of this type of surgery is the considerably high percentage of artificial knees still functioning even 20 years after the surgery. During this surgical operation, it is intended to replace damaged or worn components of the knee joint with artificially produced components or implants in order to facilitate proper knee motion of the patients and reduce disability problems and severe pain caused by joint diseases, usually osteoarthritis or rheumatoid arthritis. Apart from patients with advanced osteoarthritis, knee replacement may be suggested for younger patients with damage in the knee joint or bone or some type of deformity. Partial knee replacement is suggested in cases when damage is located only in a specific compartment of the knee. The specific purpose of this type of surgery is essentially to cap the end parts of the bones of the knee joint as well as the kneecap, by means of artificial components. More specifically, during knee replacement surgery, the main knee parts that are replaced are pertinent to the femoral and tibial surfaces near the knee joint as well as a part of the patella. For the components of the replacement implant, suitable metals or non-metallic materials are chosen and properly machined in order to resemble the natural shape of the knee as much as possible and allow for proper joint motion, resistance to wear and corrosion and biocompatibility. The machined components should adhere to highly restrictive regulations and exhibit considerable durability [2–6].

Knee replacement surgery and the design of implant components is currently considerably important and constitutes a part of ongoing research in the medical and engineering scientific communities. However, the first reported attempts to create and use artificial knee components in an actual knee replacement surgery date back to the end of the 19th century, when Gluck created the first artificial joint by ivory [7]. From that period, research is concentrated on the improvement of knee replacement components with a view to decrease considerably their failure rate and replicate proper joint motion as close as possible to the actual. In the 1950s, several new designs, such as the Waldius [8] and GUEPAR [9] knee implants were created; their basic characteristic was the bending-extending capabilities of the joint. During the next two decades, significant advances regarding the knee implant design were observed but still, artificial knee designs such as Geomedic [10] and Geometric [11] suffered from the inability to perform rotational motion and other types of knees such as Marmor [12] and Gunston [13] were exhibiting premature failure due to high contact stresses and material overloading. The ICLH (Imperial College-London Hospital) knee implant design, which was introduced by Freeman and Swanson in 1971 [14], exhibited a lower deformation and wear rate. Other designs presented during the 1970s include Total Condylar and Townley [15], which involved also the kneecap (patella) replacement and Oxford [16] and New Jersey LCS (Low Contact Stress) [17] knee designs, which used mobile bearings. In particular, the latter designs proved to possess capabilities, such as greater mobility and improved compatibility as they employed a secondary moving support surface.

Later, during the 1990s, designs employing moving bearings were also developed which were further enhanced by the use of newly developed materials and alloys and assistive guidelines from previous designs. A representative example of knee implant designs of this period is the B-P (Buechel-Pappas) Mark V which was created from Ti alloy with TiN coating or alternatively, from Co–Cr alloy in order to improve its wear resistance [18]. During this period, Walker et al. [19] introduced a new knee simulating machine with a view to test the kinematics of total knee implants, as well as the wear of these implants. Their machine was able to perform in a more realistic way than its predecessor models, which exhibited several disadvantages such as reduced accuracy in representing the forces exerted on various components of the knee implant or inadequate constraints. More recently, Harrysson et al. [20] proposed a new design method for knee implants, based on patient-specific Computed Tomography (CT) data which can provide among other, more accurate replication of the actual geometries of knee components and the reduced possibility of implant loosening. Lee et al. [21] also employed CT data in order to design the femoral component and afterwards, they created femoral components using rapid prototyping and Computer Numerical Control (CNC) machining methods. Finally, Song et al. [22] presented a thorough work related to the rapid manufacturing of femoral component using Selective Laser Melting (SLM) method which can provide a reliable way to produce customized implants. Results from all the aforementioned works can lead to useful guidelines regarding the design of knee implants. For example, it is crucial to avoid including redundant kinematic constraints for the knee implant components, to enable normal knee motion, to include moving bearings to the design, and to plan the rather copious machining process of the geometrically complex implant in an effective way.

In the present work, the methodology of designing and machining of the femoral component of total knee replacement using a 3-axis Computer Numerical Control (CNC) machining center is presented in several steps and is discussed in detail. After the machining processes take place, an evaluation of machined surface is performed by means of surface roughness measurements in order to determine the optimum process parameters for the machining of various parts of the implant. The present work is related to previous works by the members of the same scientific group, including studies on machining of femoral heads [23,24] and extends the preliminary work on the machining of knee implant, as reported in [25].

## 2. Design of the Implant Geometry and Machining Processes

### 2.1. Design of the Knee Implant Components

Despite the fact that the main focus of the present work is set on the design and machining process of the femoral component, it is considered to be crucial to study the design of the whole knee implant at the initial stage in order to determine the appropriate dimensions of this part in relation to the total knee assembly and then study the femoral component separately. The design of the knee implant components was conducted by means of Solidworks$^{TM}$ software in which both the design and manufacturing of the knee implant is able to be studied. As with every engineering design, design constraints should be properly defined in order for the implant to comply with the desired shape of implant parts and allow for normal motion of the implant components so as to provide a satisfying replacement of the damaged knee.

The design of the knee implant parts complies with the international standard ISO 7207-1 [26], in which the geometry of both total and partial knee implant components is defined in detail. More specifically, for the present work, it is assumed that the total knee implant, which is designed, corresponds to the left knee of a male human with constrained rotational movement. In Figure 1, a schematic of the bones near the knee joint is presented, along with a schematic indicating the position of the implant after the surgery. The knee implant components, which will be designed as an assembly, include the femoral component, the tibial component, the tibial articulating surface, and the patellar component, as can be seen in Figure 2.

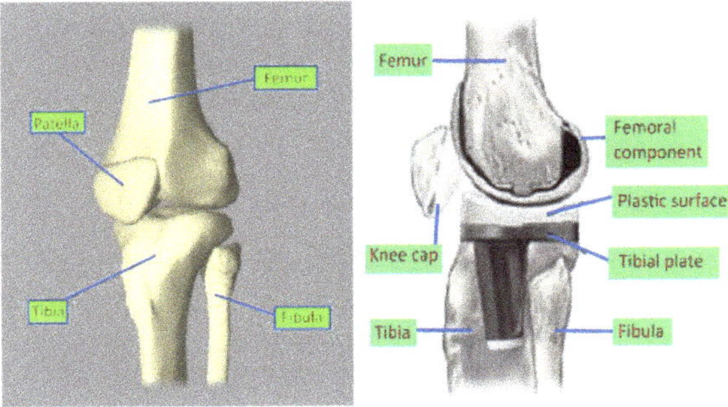


**Figure 1.** Bones around the knee joint and position of knee replacement after the knee arthroplasty.

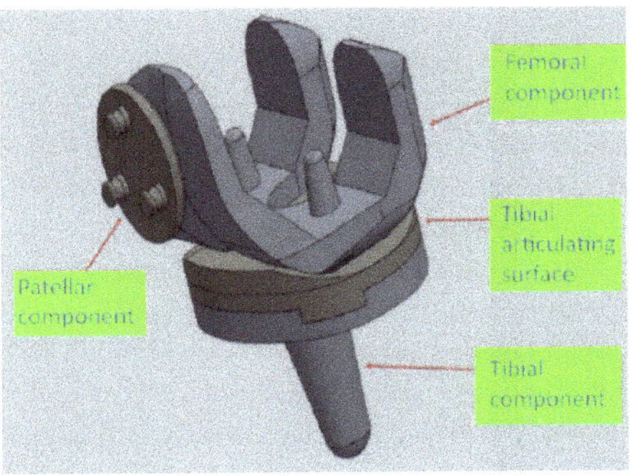


**Figure 2.** Parts of the designed knee replacement assembly.

The design of the femoral component contains two different steps, namely the design of its outer and inner surface. This approach is required due to the considerably different geometries of the outer and inner surfaces. More specifically, the outer surface is essentially a sculptured surface that is composed of various curved areas, as this shape is necessary for the contact between the femoral component and the tibial and patellar components of the implant and facilitates normal knee motion. In the outer surface, a groove is also formed, resembling the trochlear groove of the actual human femur bone. As for the design of the inner surface, it is required that it should be strongly fixed on the bone; it is usually achieved by the use of acrylic cement between the implant and the bone during the surgery. The fixation of this surface to the bone is further enhanced by two conical stems on the inner surface, intended to be inserted into two holes in the bone. Although the design of the inner surface is generally simpler, there is an important restriction regarding the thickness of the inner surfaces, as they should have the least possible thickness in order to reduce the need of material removal from the bone during the insertion of the implant.

The design of the tibial component includes the design of two different structural elements, a stem and a platform. The stem is intended to aid to the fixation of the tibial component to the tibia and the platform is essential for the connection of the tibial component to the tibial articulating surface. For the stem, it is required that its thickness is large enough to withstand the relatively high forces will be exerted on it whereas the dimensions of the platform, namely its width and depth should closely match the actual dimensions of the upper part of human tibia. As for the design of the tibial articulating surface it is subjected to two main requirements, namely to withstand the loading from the femur and implement the fixation on the tibial component. In order to fulfill these requirements, the size of the contact surface with the tibial components is designed sufficiently large so as to protect the joint from receiving excessive loading and enable normal knee motion.

Finally, the design of the patellar component aims at the facilitation of the movement of the patella on the trochlear groove of the femur. Thus, a dome-like shape for the lower surface of the patellar component is adopted in order to be able to assist to the sliding movement of this component in the trochlear groove. As for the upper surface of the patellar component, its geometrical shape is properly adjusted so that it provides strong fixation on the patella bone.

*2.2. Design of the Machining Process of the Femoral Component*

After the components of the total knee implant were carefully designed according to the international standards and relevant requirements, the focus is set on the machining process of the femoral component of the knee implant. The design of the machining process of this component, which includes complex geometrical shapes, is essential to be conducted on a specialized Computer Aided Manufacturing (CAM) software, such as SolidCAM, which can be accessed through SolidWorks software.

Machining of complex, "sculptured" surfaces is important in various industries such as automotive, aerospace and optical components industry, as well as bioengineering. One of the main challenges concerning the machining of such surfaces is the reduction of machining time, as it is considerably difficult to achieve the required dimensional accuracy and surface quality. Apart from the use of cutting tools with special geometry, the machining strategy needs to be carefully planned in order to achieve the desired geometrical features and suitable process parameters are required to be determined as well. Thus, all of these tasks need to be appropriately addressed by the use of CAM software in order to perform the machining process of the femoral component.

CAM software are specialized to assist in various stages of product manufacturing. Most commonly, they involve the use of an integrated Computer Aided Design (CAD) editor or, such as the case of SolidCAM software, are themselves integrated in the framework of a CAD software and their purpose is eventually to produce the code (G-Code) in order to control CNC machine tools for manufacturing the desired parts. Using this type of software, details about the machining process such as the definition of appropriate cutting tools for the machining process and their characteristics, the desired operations on the workpiece, such as hole drilling, contour, etc., process parameters during the various stages of the process and machining strategies can be input at first. Then, G-code can be generated, in respect to the type of CNC machine used, given that an appropriate post-processor exists. Furthermore, a simulation of the intended machining operations can be performed in order to verify that the machining will be performed safely and according to the desired goals or detect possible mistakes and perform the necessary adjustments.

The basic challenge which exists in the present work is that, due to the fact that the outer and inner surface of the femoral component contain sculptured surfaces, mounting of the workpiece on the machining center bed is difficult to be performed. For that reason, it is considered more appropriate to perform the machining process in two separate phases; during the first phase, it is intended to create the internal surface of the femoral component and during the second phase, the workpiece will be reverted, and then, machining of the outer surface will occur. Thus, in the CAM software, different coordinate systems will be defined for the two stages as well as different Stock and Target materials.

### 2.2.1. First Phase of Machining

During the first phase of machining, the emphasis is set on the creation of the inner surfaces. This phase constitutes the main material removal phase and the cutting tools will remove material up to a specific height until the second phase will take place. For the first phase, four cutting tools, both flat and ball end, will be employed, as can be also seen in Table 1. It is important to note that, in the early stages of the machining process, tools with flat end and a larger diameter are preferred in order to remove large bulk of material quickly (roughing stage) without requirements for high accuracy of produced shape. However, at the final stages of the machining process, cutting tools with smaller diameter and ball end are selected in order to render appropriately the final sculptured surface according to the desired dimensional and geometrical requirements (finishing stage). The initial dimensions of the cylindrical workpiece are: 50 mm diameter and 39.3 mm height. The workpiece material, which was chosen is stainless steel 316L, which is appropriate for the femoral component [27,28].

**Table 1.** Characteristics of the cutting tools employed in the present work.

| No | Type | Diameter (mm) | Material |
|----|------|---------------|----------|
| 1 | Flat end | 16 | carbide |
| 2 | Flat end | 6 | cobalt alloy |
| 3 | Ball end | 6 | cobalt alloy |
| 4 | Ball end | 4 | carbide |

In order to achieve the creation of the desired shape of the inner surface of the femoral component, the 16 mm diameter cutting tool was first employed in order to reduce the initial height of the workpiece at several passes, as can be seen in Figure 3a. Afterwards, a cavity was formed in the region where the inner surfaces of the implant will later be created. Using the same cutting tool, a contour cutting process was also performed in order to create the front and back sculptured surfaces. The next stage, as depicted in Figure 3b, involved the use of 6 mm flat end cutting tool in order to continue the contour cutting process with the rendering of more features of the implant surface, and then, the 6 mm ball end tool was employed to create the final shape of the front sculptured surface. Finally, the finishing stage was implemented by using the 4 mm diameter cutting tool both for the front and the back sculptured surfaces, as can be seen in Figure 3c.

### 2.2.2. Second Phase of Machining

The second phase of machining is related to the machining of the outer surface of the femoral component. For this phase of machining, three cutting tools are selected, both the flat and ball end. Initially, as can be seen in Figure 4a,b, a 16 mm flat end cutting tool is used to perform contour cutting at a fixed height at each pass, and then, a 6 mm flat end cutting tool performed cutting with fixed y coordinate at each pass (roughing phase). In the end, the finishing process was implemented using a 4 mm ball end cutting tool using the same strategy as with the 6 mm diameter cutting tool, as can be seen in Figure 4c.

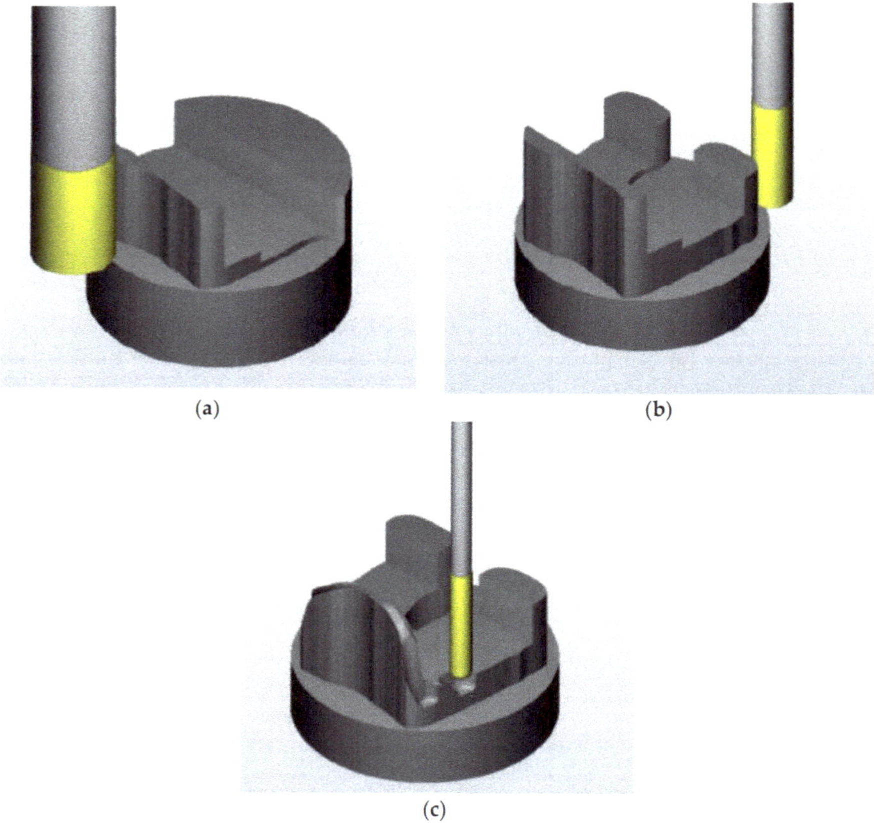

(a)      (b)

(c)

**Figure 3.** Snapshots from the simulation of the first machining stage: (**a**) Machining of the front surface of the femoral component using 16 mm diameter tool; (**b**) Machining of the back surface of the femoral component using 6 mm diameter tool; and, (**c**) Machining of inner surfaces of the femoral component using 4 mm ball end tool.

*2.3. Initial Machining Test*

After the definition of machining operations and production of G-code in the CAM software and after the simulation of the machining processes was successfully finished, it was decided that a test run, replicating the machining process in the actual CNC machining, was necessary before the final machining process in order to ensure that the generated G-code was producing an accurate and reliable outcome and verify the simulation results. Furthermore, it is important to test the behavior of the cutting tools during the machining process in order to choose the appropriate process parameters that lead to avoidance of chattering, as it is impossible to be determined from the simulation in the CAM software. The test run was performed on a cylindrical bulk of polymer material, and was concluded successfully, indicating that the actual machining process can be performed without problems.

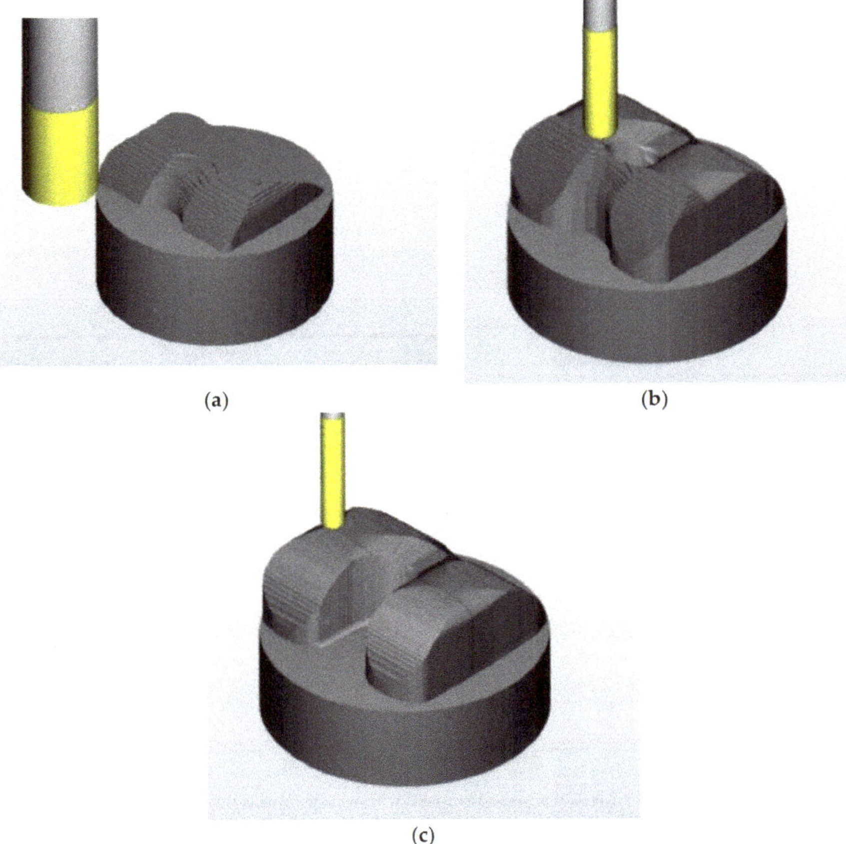

**Figure 4.** Snapshots from the simulation of the second machining stage: (**a**) Roughing stage of the second machining phase using 16 mm diameter tool; (**b**) Machining of sculptured surface of the femoral component using 6 mm ball end tool; and, (**c**) Finishing stage of the second machining phase, using 4 mm diameter ball end tool.

## 3. Machining of the Implant and Measurement of Surface Roughness

### 3.1. Machining Process of the Implant

After the various stages of the machining process were designed in the CAM software and successfully verified by the test run, the setup of the 3-axis CNC machining center was performed. As the initial bulk is a cylindrical workpiece, it was fixed on the machine with a chuck, as depicted in Figure 5.

After the cutting tools were selected and the necessary setup was performed on the CNC machine, the first phase of the machining process took place. The first phase was expected to last for a relatively long time, as it is required to select small depths of cut and low cutting speed when machining stainless steel workpieces. After the machining process was completed, the inner surface of the implant was created, as can be seen in Figure 6. It is worth noting that the existence of several markings on the produced surfaces was observed, created by the contact of the upper part of the cutting tool with these surfaces when material removal was performed on the lower parts of the inner surface of the implant.

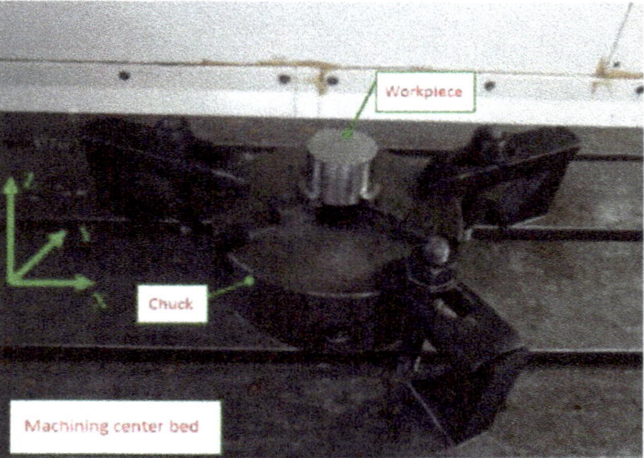


**Figure 5.** Clamping of workpiece on the machining center bed.

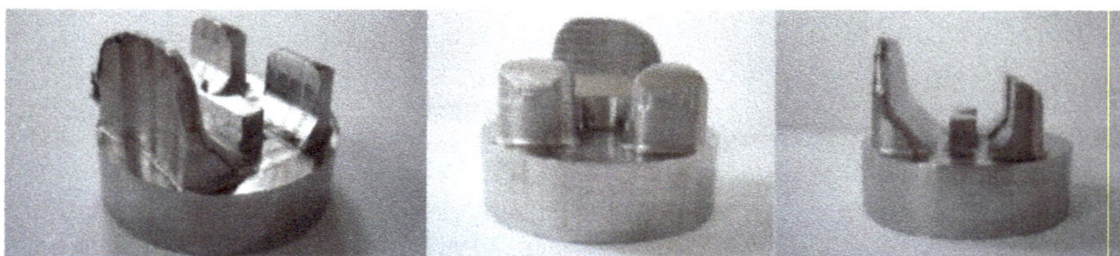

**Figure 6.** Views of the workpiece after the first machining stage.

During the second phase of the machining process the workpiece is mounted on the machine tool bed after it is reverted, by using a specially designed platform, as can be seen in Figure 7. The use of this platform is required as it is impossible to fix otherwise the workpiece appropriately on the machine tool. In order to ensure the stability of fixation of the implant, it is required that the internal surfaces of the workpiece are properly aligned on the platform surfaces, parallel to the machining center bed. After the implant is mounted on the platform and fixation is properly performed by two screws, the platform is clamped on a chuck.

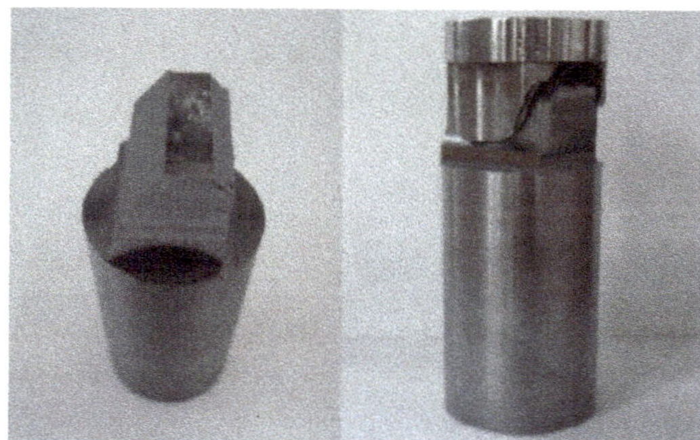

**Figure 7.** The platform for mounting the workpiece on the machining center bed for the second machining stage.

When the aforementioned procedure for the machining of the outer surface of the implant is completed, as can be seen in Figure 8. This process was repeated two times with different process parameters at various zones of the implant during finishing stage, in order to investigate the effect of process parameters on surface quality. The process parameters that were employed for the machining of all three implant during the finishing stage are presented in Table 2.

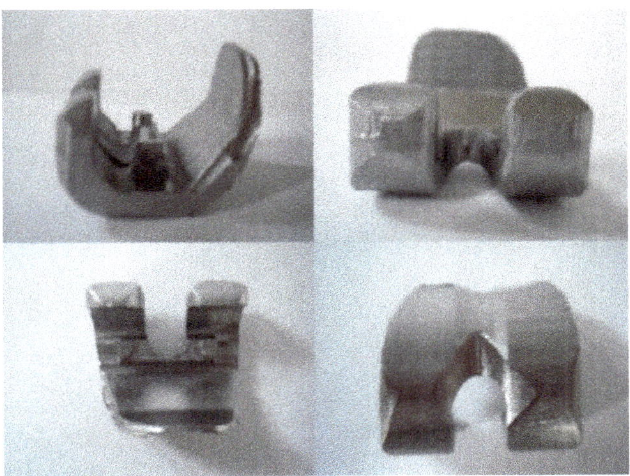

**Figure 8.** Views of the implant after the second machining stage.

**Table 2.** Machining parameters values for each zone and each component.

| Component | | Zones of the Component | | | | |
|---|---|---|---|---|---|---|
| | | 1 | 2 | 3 | 4 | 5 |
| 1st | Feed rate (mm/min) | 100 | 100 | 100 | 100 | 100 |
| | Spindle speed (rpm) | 2500 | 2500 | 2500 | 2500 | 2500 |
| 2nd | Feed rate (mm/min) | 90 | 70 | 50 | 100 | 90 |
| | Spindle speed (rpm) | 2500 | 2500 | 2500 | 2250 | 2250 |
| 3rd | Feed rate (mm/min) | 70 | 50 | 100 | 90 | 70 |
| | Spindle speed (rpm) | 2250 | 2250 | 2000 | 2000 | 2000 |

## 3.2. Surface Quality Evaluation

After the machining process of the femoral component is completed, it is considered important to evaluate the quality of the produced surfaces, as it is directly connected to the tribological behavior and wear resistance of the implant. Inappropriate machining conditions, leading to excessive surface roughness, can prevent not only the adequate performance and reliability of the implant but also its life cycle [29]. Furthermore sufficient surface quality after machining of the implant reduces the need for further processes, such as polishing. For that reason, surface roughness measurements were conducted on the three different implants with a view to determine the optimum process parameters for the finishing stage of the implant machining process. Due to the fact that the implant contains sculptured surfaces, the measurement of the surface roughness with a profilometer is a demanding process and special care has to be paid for the positioning of the measuring device and also for correct sampling length ($L_n$) and cut-off length ($L_c$) values. In the present work, a Surtronic 3+ Taylor Hobson profilometer was employed and a sampling length ($L_n$) of 2.4 mm was selected, as well as a cut-off length ($L_c$) of 0.8 mm. As per the manufacturer manual, the used profilometer has a resolution of 0.01 μm and accuracy of parameters is given as 2% of reading plus least significant digit in μm. Surface roughness was performed in each of the five zones presented in Figure 9, both in the left and the right side (or inner and outer side) of the upper surface of the implant (the region of the lateral and medial condyles, respectively) and the measurements are repeated three times. In the following analysis the lateral and medial condyles are referred to as inner and outer sides, respectively.

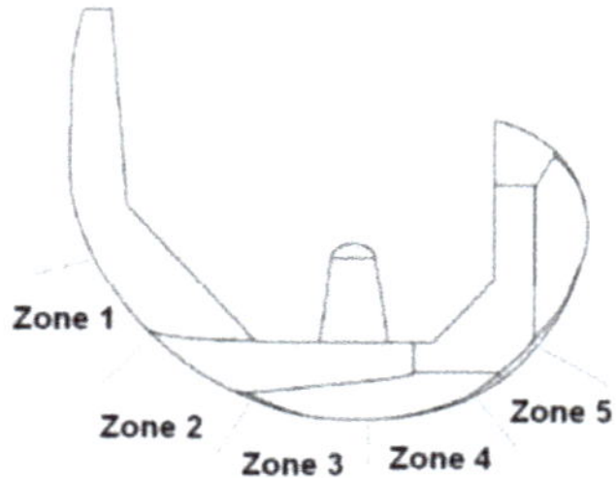


**Figure 9.** Zones of surface roughness measurement.

The evaluation of surface roughness was performed by Ra, Rq, Rt, which represent the arithmetic average surface roughness, root mean squared roughness, and maximum height of the profile, respectively. Especially, Ra is the most popular indicator of surface roughness in industrial practice until today. As for the first femoral component, which was machined with the same conditions in every zone, it can be seen from Figure 10a,b, that there are some differences in the Ra values in the different zones.

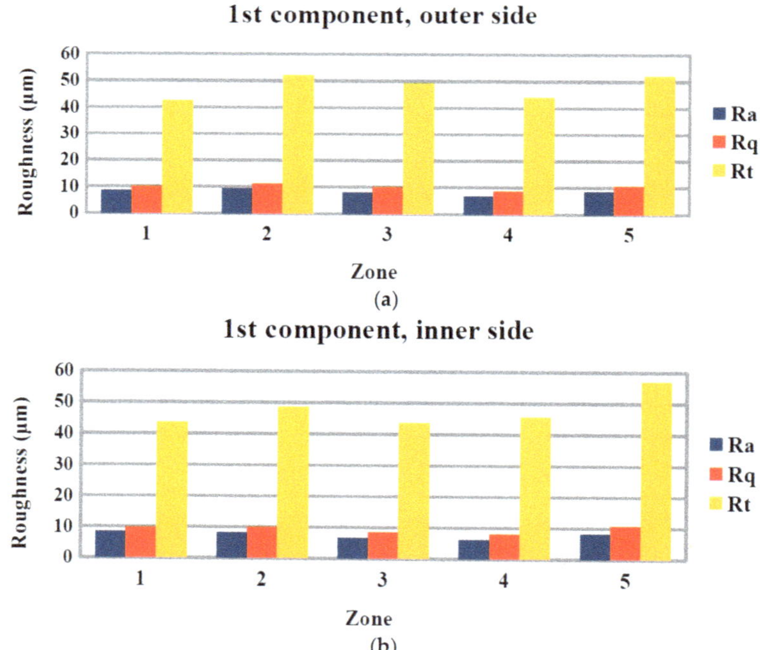


**Figure 10.** Surface roughness measurement results for the first femoral component for (**a**) outer and (**b**) inner side.

These differences can be directly attributed to the different geometry of each zone, as zones 1, 2, and 5 are more curved, whereas region 3 and 4 are almost flat. Thus, it is observed that the finishing process is more effective in the flat regions with values of Ra of about 6–7 µm, whereas Ra exceeds 8 µm in the other regions. Similar trends can be observed in the case of Rq and Rt measurement with the exception of zone 1, which has a relatively low value of Rt when compared to the other zones.

A comparison of surface roughness measurements between the different components, machined with different conditions, can reveal the effect of process parameters to the surface quality of each zone of the femoral component. For zone 1, comparison between results depicted in Figures 10–12 show that, a reduction of feed rate resulted in an increase of Ra and subsequent reduction of both feed rate and spindle speed resulted in a further small increase of Ra. For zone 2, the change of process parameters resulted in almost unchanged values of Ra. For zone 3, a decrease of feed rate resulted in a considerable decrease of average surface roughness, whereas a decrease of spindle speed resulted in an increase of Ra. For zone 4, a decrease of spindle speed resulted in an increase of Ra, whereas a further increase of spindle speed with a slight decrease of feed rate led to a slight decrease of Ra. Finally, for zone 5, it was observed that the reduction of both feed rate and spindle speed resulted initially in higher Ra and a subsequent reduction of both the parameters resulted in slightly lower Ra.

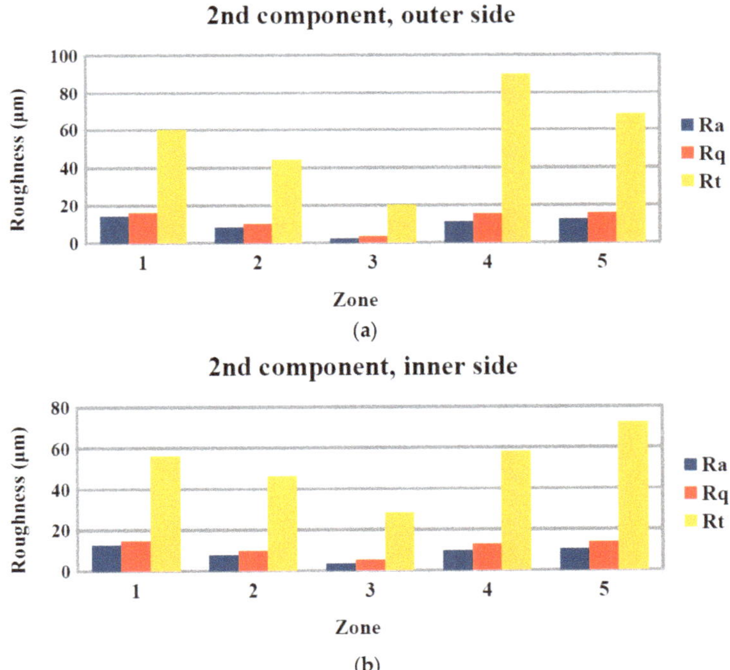


**Figure 11.** Surface roughness measurement results for the second femoral component for (**a**) outer and (**b**) inner side.

As usually, the effect of feed rate on Ra is that a decrease of feed rate leads to a decrease of Ra and also an increase of spindle speed is beneficial to the surface quality, the results for some of the five zones seem somewhat unexpected. However, these discrepancies can be attributed to the different curvature of each zone; results consistent with the aforementioned behavior are exhibited in zones which are fairly flat, such as zones 3 and 4, but in zones that are more curved different trends exist. The same trends are observed in the case of Rq and Rt measurements on the three femoral components.

The difference between measurements in the inner and the outer side of the upper surface of the femoral component was also investigated during the analysis of surface roughness of the implants.

Regarding Ra values, it was found that, with the exception of one zone for the 2nd and 3rd component and two zones for the 1st, the general trend of variations of Ra in respect to different geometry and process parameters was similar. Furthermore, it was observed Ra values were larger in the vast majority of measurements in the outer side rather than the inner side.

Finally, the previous analysis allowed for the determination of optimum process parameters, namely feed rate and spindle speed for the finishing stage. Within the examined range of process parameters values the optimum values for each zone regarding Ra, were observed in the first component for zones 1, 4, 5, and for the second component for zones 2 and 3. Similar conclusions were drawn when examining the values of Rq and Rt.

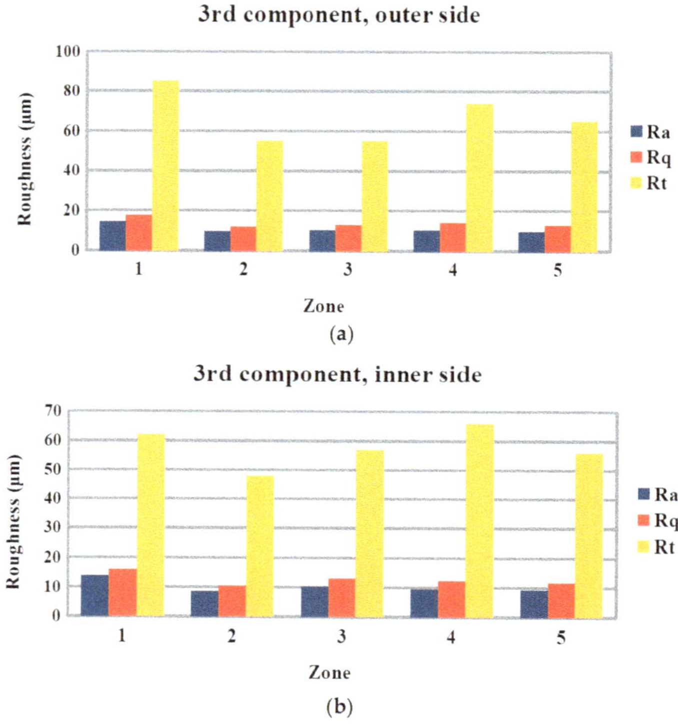


**Figure 12.** Surface roughness measurement results for the third femoral component for (**a**) outer and (**b**) inner side.

## 4. Conclusions

In the present work, various stages of manufacturing the femoral component of knee prosthesis are presented, including the design of geometry and machining operations, as well as the actual machining process and the subsequent determination of surface quality.

The specific requirements for the design of the femoral component in respect to the total knee replacement assembly were determined, and particular situations that require special attention during machining were identified. These particularities were properly taken into consideration during the design of the required machining operations using SolidCAM software.

Furthermore, after machining three different femoral components with variable process parameters values during the finishing stage several important conclusions were drawn. Especially, surface roughness was shown to vary considerably with changes in the surface curvature and the effect of process parameters to the surface quality was also shown to be dependent of the curvature of the surface. Finally, the optimum parameters, namely feed rate and spindle speed, for the reduction of surface roughness for various zones of the femoral implant were determined.

**Author Contributions:** Nikolaos I. Galanis and Dimitrios E. Manolakos conceived and designed the experiments; Nikolaos I. Galanis performed the experiments; Nikolaos I. Galanis and Nikolaos E. Karkalos analyzed the data; Angelos P. Markopoulos contributed materials and analysis tools; Nikolaos E. Karkalos and Angelos P. Markopoulos wrote the paper.

**Conflicts of Interest:** The authors declare no conflict of interest.

## References


1. Carr, B.C.; Goswami, T. Knee implants—Review of models and biomechanics. *Mater. Des.* **2009**, *30*, 398–413. [CrossRef]
2. Magdziak, M. The influence of a number of points on results of measurements of a turbine blade. *Aircr. Eng. Aerosp. Technol.* **2017**, *89*, 953–959. [CrossRef]
3. Felhő, C.; Kundrák, J. Comparison of theoretical and real surface roughness in face milling with octagonal and circular inserts. *Key Eng. Mater.* **2014**, *581*, 360–365. [CrossRef]
4. Kundrák, J.; Varga, G. Possibility of Reducing Environmental Load in Hard Machining. *Key Eng. Mater.* **2012**, *496*, 205–210. [CrossRef]
5. Kundrák, J.; Varga, G.; Deszpoth, I.; Molnar, V. Some Aspects of the Hard Machining of Bore Holes. *Appl. Mech. Mater.* **2013**, *309*, 126–132. [CrossRef]
6. Illés, B.; Tamás, P.; Dobos, P.; Skapinyecz, R. New challenges for quality assurance of manufacturing processes in industry 4.0. *Solid State Phenom.* **2017**, *261*, 481–486. [CrossRef]
7. Brand, R.A.; Mont, M.A.; Manring, M.M. Biographical sketch: Themistocles Gluck (1853-1942). *Clin. Orthop. Relat. Res.* **2011**, *469*, 1525–1527. [CrossRef] [PubMed]
8. Wilson, F.C.; Fajgenbaum, D.M.; Venters, G.C. Results of knee replacement with the Walldius and geometric prostheses. A comparative study. *J. Bone Jt. Surg. Am.* **1980**, *62*, 497–503. [CrossRef]
9. Aubriot, J.-H.; Deburge, A.; Genet, J.-P. GUEPAR hinge knee prosthesis. *Orthop. Traumatol. Surg. Res.* **2014**, *100*, 27–32. [CrossRef] [PubMed]
10. Van Loon, C.J.; Hu, H.P.; Van Horn, J.R.; De Waal Malefijt, M.C. The Geomedic knee prosthesis. A long-term follow-up study. *Acta Orthop. Belg.* **1993**, *59*, 40–44. [PubMed]
11. Coventry, M.B.; Finerman, G.A.; Riley, L.H.; Turner, R.H.; Upshaw, J.E. A new geometric knee for total knee arthroplasty. *Clin. Orthop. Relat. Res.* **1972**, *83*, 157–162. [CrossRef] [PubMed]
12. Kolstad, K.; Wigren, A. Marmor Knee Arthroplasty: Clinical Results and Complications during an Observation Period of at least 3 Years. *Acta Orthop. Scand.* **1982**, *53*, 651–661. [CrossRef] [PubMed]
13. Lavernia, C.; C Alcerro, J.; Contreras Raygoza, J. Knee arthroplasty: Growing trends and future problems. *Int. J. Clin. Rheumatol.* **2010**, *5*, 565–579. [CrossRef]
14. Goldberg, V.M.; Henderson, B.T. The Freeman-Swanson ICLH total knee arthroplasty. Complications and problems. *JBJS* **1980**, *62*, 1338–1344. [CrossRef]
15. Mallory, T.H.; Smalley, D.; Danyi, J. Townley Anatomic Total Knee Arthroplasty Using Total Tibial Component with Cruciate Release. *Clin. Orthop. Relat. Res.* **1982**, *169*, 197–201. [CrossRef]
16. Goodfellow, J.W.; Tibrewal, S.B.; Sherman, K.P.; O'Connor, J.J. Unicompartmental Oxford Meniscal knee arthroplasty. *J. Arthroplast.* **1987**, *2*, 1–9. [CrossRef]
17. Buechel, F.F.S.; Buechel, F.F.J.; Pappas, M.J.; Dalessio, J. Twenty-year evaluation of the New Jersey LCS Rotating Platform Knee Replacement. *J. Knee Surg.* **2002**, *15*, 84–89. [PubMed]
18. Buechel, F.F.; Pappas, M.J. *Principles of Human Joint Replacement: Design and Clinical Application*; Springer International Publishing: Berlin/Heidelberg, Germany, 2015.
19. Walker, P.S.; Blunn, G.W.; Broome, D.R.; Perry, J.; Watkins, A.; Sathasivam, S.; Dewar, M.E.; Paul, J.P. A knee simulating machine for performance evaluation of total knee replacements. *J. Biomech.* **1997**, *30*, 83–89. [CrossRef]
20. Harrysson, O.L.A.; Hosni, Y.A.; Nayfeh, J.F. Custom-designed orthopedic implants evaluated using finite element analysis of patient-specific computed tomography data: Femoral-component case study. *BMC Musculoskelet. Disord.* **2007**, *8*, 91. [CrossRef] [PubMed]

21. Lee, J.; Chen, H.; Luo, C.-W.; Chang, K.-Y. Rapid Prototyping and Multi-axis NC Machining for The Femoral Component of Knee Prosthesis. *Life Sci. J.* **2010**, *6*, 73–77.
22. Song, C.; Yang, Y.; Wang, Y.; Wang, D.; Yu, J. Research on rapid manufacturing of CoCrMo alloy femoral component based on selective laser melting. *Int. J. Adv. Manuf. Technol.* **2014**, *75*, 445–453. [CrossRef]
23. Galanis, N.I.; Manolakos, D.E. Surface roughness prediction in turning of femoral head. *Int. J. Adv. Manuf. Technol.* **2010**, *51*, 79–86. [CrossRef]
24. Galanis, N.; Manolakos, D. Surface roughness of manufactured femoral heads with high speed turning. *Int J. Manuf. Mater. Mech. Eng.* **2009**, *5*, 371–382. [CrossRef]
25. Markopoulos, A.P.; Karkalos, N.E.; Galanis, N.I.; Manolakos, D.E. Design and Machining of the Femoral Component of Total Knee Implant. *Solid State Phenom.* **2017**, *261*, 313–320. [CrossRef]
26. Denkena, B.; Köhler, J.; Turger, A.; Helmecke, P.; Correa, T.; Hurschler, C. Manufacturing conditioned wear of all-ceramic knee prostheses. *Procedia CIRP* **2013**, *5*, 179–184. [CrossRef]
27. Otani, T.; Whiteside, L.A.; White, S.E.; McCarthy, D.S. Effects of femoral component material properties on cementless fixation in total hip arthroplasty. A comparison study between carbon composite, titanium alloy, and stainless steel. *J. Arthroplast.* **1993**, *8*, 67–74. [CrossRef]
28. Gervais, B.; Vadean, A.; Raison, M.; Brochu, M. Failure analysis of a 316L stainless steel femoral orthopedic implant. *Case Stud. Eng. Fail. Anal.* **2016**, *5*, 30–38. [CrossRef]
29. Affatato, S.; Ruggiero, A.; Grillini, L.; Falcioni, S. On the roughness measurement of the knee femoral components. In Proceedings of the 10th International Conference BIOMODLORE 2013, Vilnius, Lithuania, 20–22 September 2013; Vilnius Gediminas Technical University Press Technica: Vilnius, Lithuania, 2013; pp. 16–18. [CrossRef]

*Article*

# The Dimensional Precision of Forming Windows in Bearing Cages

**Marius Rîpanu, Gheorghe Nagîţ, Laurenţiu Slătineanu and Oana Dodun ***

Department of Machine Manufacturing Technology, "Gheorghe Asachi" Technical University of Iaşi, Blvd. D. Mangeron, 59 A, Iaşi 700050, Romania; ripanumariusionut@yahoo.com (M.R.); nagit@tcm.tuiasi.ro (G.N.); slati@tcm.tuiasi.ro (L.S.)
* Correspondence: oanad@tcm.tuiasi.ro; Tel.: +40-747-144-604

Received: 5 February 2018; Accepted: 24 February 2018; Published: 1 March 2018

**Abstract:** In the case of double row tapered roller bearings, the windows found in bearing cages could be obtained using various machining methods. Some such machining methods are based on the cold forming process. There are many factors that are able to affect the machining accuracy of the windows that exist in bearing cages. On the dimensional precision of windows, the clearance between punches and die, the work stroke length, and the workpiece thickness could exert influence. To evaluate this influence, experimental research was developed taking into consideration the height and the length of the cage window and the distance between the contact elements of the cage. By mathematical processing of the experimental results, empirical mathematical models were determined and analyzed. The empirical models highlighted the intensity of the influence exerted by the considered forming process input factors on the dimensional precision of the windows obtained in bearing cages.

**Keywords:** double row tapered roller bearing; bearing cage; window forming; dimensional precision; clearance between punch and die; work stroke length; workpiece thickness; mathematical empirical model

## 1. Introduction

The rolling bearings are machine elements usually intended to support parts found in a rotation movement. Essentially, they include *rolling elements* and *rings*. To separate the rolling elements, *bearing cages* could be used, as sometimes *sealing elements* are provided to prevent the penetration of foreign objects or substances between bearing elements found in a relative movement to one another.

A category of roller bearings that are used in difficult service conditions, when there is the necessity of compensating the eventual deviations of the coaxiality of supporting bearings or the shaft bending, are the double row tapered roller bearings (Figure 1a). These roller bearings are characterized by a certain self-regulation on the raceway of the outer ring.

The bearing cages presented in the bearings have generally the role of preventing the contact between rolling elements (rolls or balls). There are various methods of manufacturing the bearing cages, considering their shapes and materials. Thus, for example, in the case of cylindrical rolls, the bearing cages could be achieved essentially by successive processes of drawing, perforating the central hole, and cutting the windows. As workpieces, metallic plates that have circular shapes could be taken into consideration before drawing.

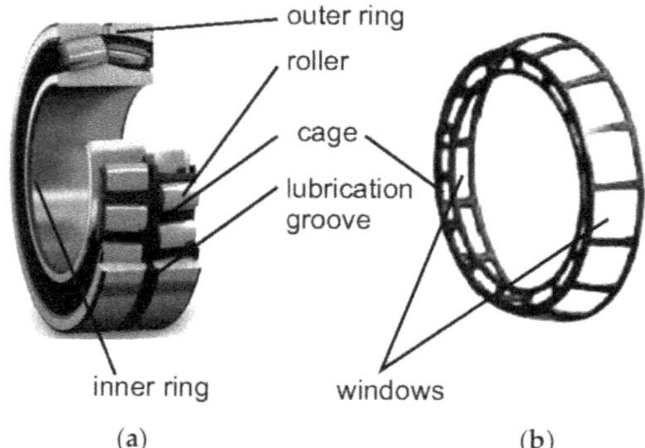


**Figure 1.** (**a**) Double row tapered roller bearing; (**b**) details of bearing cage.

The main requirements for bearing cages refer to their accuracy, their surface roughness, and material resistance to wear processes.

If the cage machining accuracy is analyzed, in addition to the accuracy of the overall dimensions, the precision of the dimensions corresponding to the windows in which the rolling elements could be placed is important. The concept of *dimensional precision* refers to the deviation of the real dimensions from their nominal values.

Over the years, the results of certain research concerning the accuracy and the surface state of bearing cages or generally of the windows obtained in sheet type workpieces by forming methods were published.

Thus, Kibe et al. considered that the general position of the punch in the die could exert a significant influence on the machining accuracy in the case of shearing processes [1]. They proposed a method that was able to increase the precision of measuring and adjusting the position of the punching tool, especially when processing metallic thin sheets. The experiments were achieved on workpieces made of phosphor bronze and aluminum that had a thickness of 0.2 mm. The researchers highlighted the influence exerted by the clearance between punch and die on the machining accuracy, taking into consideration the shape of the cross-section and the diameter of the hole obtained by punching.

Subsequently, Kibe and Mitsui investigated the influence exerted by the misalignment of the punch and die positions on the cross-section diameter of the hole and on the out of the roundness in the case of punching applied to workpieces made of phosphor bronze [2].

Istiawan and Mahardika studied the influence exerted by the clearance and the punch speed on the surface quality evaluated by means of burnish area, fracture, and burr zone, in the case of punching a hole with the diameter of 800 μm in a workpiece with a thickness of 300 μm [3]. They noticed the low influence exerted by the punch speed on the surface quality.

Ardeshana and Mehta took into consideration the problem of designing a punch and a die for manufacturing of a bearing cage specific to a taper roller bearing [4]. They considered that the simultaneous cutting of all the windows in the bearing cage could contribute significantly to the decrease of the production time, and thus to the increase of the productivity.

Zhao et al. considered that an improvement of the precision manufacturing of bearing cages by the stamping process could be possible by using the reliability theory to adequately establish the thickness of the plate used as the workpiece in the process of the manufacture of the bearing cage [5]. They modeled the behavior of the bearing cage during the punching process by using the ANSYS finite element analysis.

Various solutions were also considered in patents that referred to the producing windows in roller bearings cages by distinct manufacturing processes [6–10]. Within these patents, only the processes and equipment for obtaining windows in bearing cages were discussed, without offering details about the dimensional precision of the cages' windows that were obtainable by such manufacturing processes. Even the roller bearings are not complex systems; they could be incorporated in such systems, and principles corresponding to the mechanical systems could be taken into consideration when analyzing the roller bearing and its manufacture [11–15].

The main aim of the research presented in this work was to highlight the influence exerted by the workpiece thickness, punch work stroke, side clearance, and axial clearance between punch and die on the precision of the main dimensions of the windows that exist in the bearings cages in the case of the double row tapered roller bearings. Empirical mathematical models that were able to offer information concerning the intensity of the influences of the abovementioned forming process input factors on three main dimensions of the bearing cage windows were determined.

## 2. Theoretical Analysis

In order to achieve the windows in the bearing cage from Figure 1b, a stamping process is applied. On could consider that the stamping process develops on hydraulic pressing equipment (Figure 2a) and using a mold that supposes the axial movement of a central conical mandrel. Due to this movement and to the conical shape of the central conical mandrel, the punches that have an approximate L shape are forced to radially move, materializing the process of window cutting (Figure 2b,c). Since there is a number of punches that is equal to the number of windows that exist in the bearing cage, at a single work stroke of the central conical mandrel, all of the windows are cut.

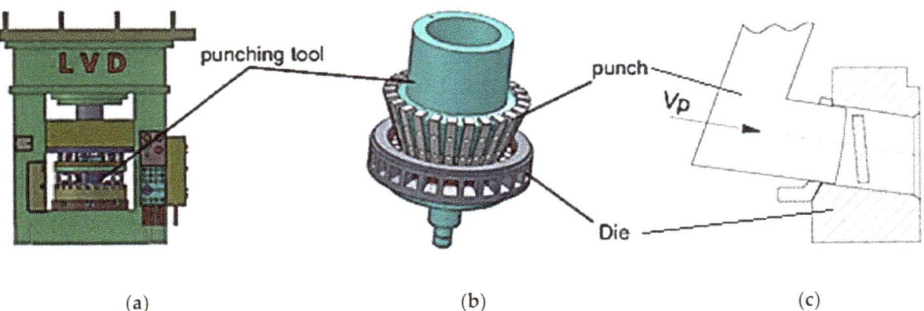


**Figure 2.** Achieving the windows in bearing cages: (**a**) LVD 600 hydraulic pressing equipment used for cutting the windows in bearings cages; (**b**) components of the cutting tool; (**c**) cutting of a window.

In Figure 3, the shape and the main dimensions specific to a window that exists in the bearing cage could be seen. The main dimensions of the window are the window width $W_w$, the window height $H_w$, and the distance $D_{rt}$ between the retention thresholds belonging to the same window. As a consequence of the punch movement, a part of the workpiece material is detached, and a hole is thus generated in the bearing cage (Figure 4).

During the rolling bearing service, especially the window height $H_w$, and the distance $D_{rt}$ between retention thresholds are important, since they determine the clearance between the bearing rolls and the bearing cage. These dimensions are determined during the rolling bearing design activity by taking into consideration the recommended or imposed values for the length and the maximum diameter of the bullet roll, the axial, the side, and the tangential clearance between the roll and the bearing cage.

There are many factors that are able to exert influence on the precision of the dimensions $H_w$, $W_w$, and $D_{rt}$ during the process of window cutting in accordance with the machining scheme presented in Figure 2b.

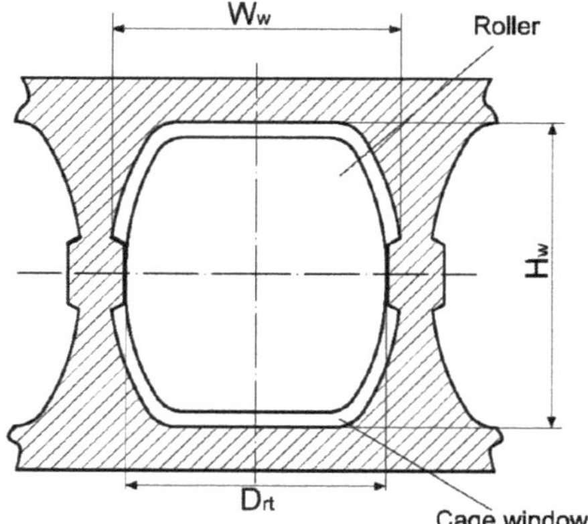

**Figure 3.** Shape and main dimensions of the window existing in the bearing cage.

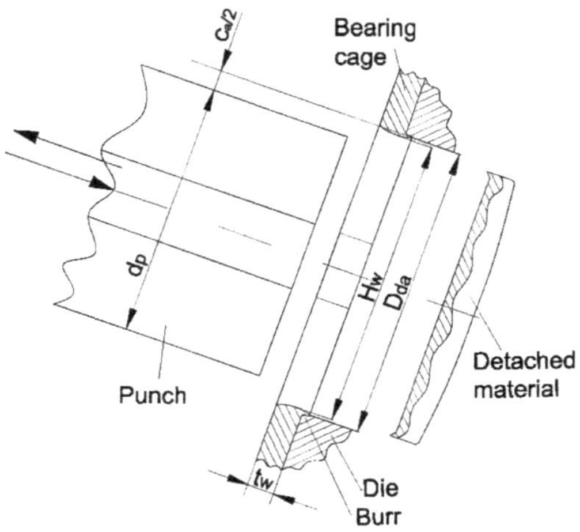


**Figure 4.** Window in the bearing cage after withdrawal of the punch.

The factors that are able to influence the precision of the window cutting in the bearing cage could be grouped in the following way:

1. Geometry of punch and die active zones;
2. Size of clearance between punch and die;
3. Length of punch work stroke and cutting speed;
4. Certain physical–mechanical properties of workpiece material (friction coefficient, shear resistance, hardness, yield strength, formability, etc.);
5. Level of wear that characterizes the active zones of the punch and die, etc.

The clearance between the punch and the die could be determined as a half of the difference between the dimension of the hole that exists in the die and the conjugate dimension of the punch, both in cross-sections. For example, in the case of windows height $H_w$, the relation for determining the axial clearance $c_a$ (clearance measured along the roller axis) is the following:

$$C_a = \frac{D_{da} - d_{pa}}{2},$$

(1)

in which $D_{da}$ is the axial dimension of the cross section of the hole existing in the die, and $d_{pa}$ is the adequate dimension of the punch cross section.

As hypothesized, one can consider that the increase of the clearance between the active zones of punches and die and wear level of these main components will determine the decrease of the cutting process accuracy and of the burr height.

On could suppose a monotone influence exerted by the clearance $c$ between punch and die, by the workpiece thickness $t_w$ and by the length of punch work stroke $l_s$ on the accuracy of the dimensions that characterize the window cut in the bearing cage. Accepting such a hypothesis, one could consider that a power type equation could be able to highlight the intensity of the influence exerted by the workpiece thickness $t_w$, the length $l_s$ of work stroke, the side clearance $c_s$, and axial clearance $c_a$ on the deviation $\Delta D_i$ of the investigated dimension $D_i$ from its nominal value:

$$\Delta D_i = C_0 \cdot t_w^{C_1} \cdot l_s^{C_2} \cdot c_a^{C_3} \cdot c_s^{C_4},$$

(2)

in which the coefficient $C_0$ and the exponents $C_1$, $C_2$, $C_3$, and $C_4$ could be experimentally determined. The values of the exponents $C_1$, $C_2$, $C_3$, and $C_4$ will characterize the intensity of the influence exerted by the considered process input factors on the cutting accuracy of the bearing cage window.

## 3. Materials and Methods

In order to test the hypotheses concerning the influence exerted by the considered input factors of the window cutting process on window cutting accuracy, experimental research was designed and carried out [16,17].

The experiments were made on the LVD 600 type hydraulic pressing equipment (Figure 2a). The punch and the die correspond to the schematic representation from Figure 2b.

As material of the bearing cage, the steel DD13 (SR EN 10111: 2008) was considered. In the chemical composition of this steel, there are the following main components: 0.034% C, 0.015% Si, 0.25% Mn, 0.02% P, 0.0130% S, etc. The tensile strength of this steel is about 400 MPa.

The possibility was considered of developing an experimental research in which the results could be mathematically processed by means of software based on the method of last squares so that, finally, empirical mathematical models could be determined and analyzed. These empirical mathematical models have to highlight the intensity of the influence exerted by considered process input factors on the sizes of the parameters of technological interest, from the point of view of cutting accuracy.

Regarding cutting process input factors (independent variables), the followings were considered: the workpiece thickness $t_w$, the length of the punch work stroke $l_s$, the side clearance $c_s$ (clearance measured in a plane perpendicular on the axis of the virtual roller), and the axial clearance $c_a$ between the punch and die.

The values of the process input factors $t_w$, $l_s$, $c_l$, and $c_a$, and of the output parameters (deviation $\Delta H_w$ of the window height $H_w$, deviation $\Delta W_w$ of the window width $W_w$, and deviation $\Delta D_{rt}$ of the distance $D_{rt}$ between the cage retainer thresholds, were measured using the MarSurf XC 2 contour measuring station. The station guide deviation is lower than 1 μm/120 mm.

As experimental result, the deviations of the measured dimensions from their nominal values were considered.

The experimental conditions and results were presented in Table 1. In the column nos. 2, 3, 4, and 5, the values of the process input factors were included, while in the columns nos. 6, 7, and 8, the sizes of the process output parameters were mentioned.

Table 1. Experimental conditions and results.

| | Values of the Process Input Factors | | | | Values of the Process Output Parameters | | |
|---|---|---|---|---|---|---|---|
| Exp. No. | Workpiece Thickness, $t_w$, mm | Stroke Length, $l_s$, mm | Side Clearance $c_l$, mm | Axial Clearance, $c_a$, mm | Deviation of the Window Height, $\Delta H_w$, mm | Deviation of Window Width, $\Delta W_w$, mm | Deviation of Distance between the Retention Thresholds, $\Delta D_t$, mm |
| Column No. 1 | 2 | 3 | 4 | 5 | 6 | 7 | 8 |
| 1 | 3.5 | 5.5 | 0.18 | 0.58 | 0.2061 | 0.0782 | 0.0993 |
| 2 | 3.5 | 6.5 | 0.11 | 0.51 | 0.2089 | 0.0684 | 0.0931 |
| 3 | 3.5 | 7.5 | 0.11 | 0.51 | 0.2141 | 0.0697 | 0.0941 |
| 4 | 3.5 | 7.5 | 0.18 | 0.58 | 0.2685 | 0.0879 | 0.1101 |
| 5 | 4 | 5.5 | 0.11 | 0.51 | 0.0797 | 0.0551 | 0.0804 |
| 6 | 4 | 5.5 | 0.18 | 0.58 | 0.1521 | 0.0676 | 0.0918 |
| 7 | 4 | 7.5 | 0.11 | 0.51 | 0.1989 | 0.0642 | 0.0884 |
| 8 | 4 | 7.5 | 0.18 | 0.58 | 0.2512 | 0.0766 | 0.0995 |
| 9 | 5 | 5.5 | 0.18 | 0.58 | 0.1478 | 0.0660 | 0.0908 |
| 10 | 5 | 6.5 | 0.11 | 0.51 | 0.1048 | 0.0590 | 0.0842 |
| 11 | 5 | 7.5 | 0.11 | 0.51 | 0.1674 | 0.0614 | 0.0862 |
| 12 | 5 | 7.5 | 0.18 | 0.58 | 0.2475 | 0.0755 | 0.0995 |

## 4. Results

The experimental results were mathematically processed by means of software based on the use of the method of least squares [18]. The software allows for the consideration of five distinct mathematical empirical models. The adequacy of these empirical models for the experimental results could be estimated by means of the so-called Gauss's criterion. The value of the Gauss's criterion could be determined as a sum of squares of the differences between the measured values and the values determined using the empirical mathematical models, for the same experimental points.

As results of experimental mathematical processing, the following mathematical empirical relations were determined:

$$\Delta H_w = 0.0311 \cdot t_w{}^{-0,877} \cdot l_s{}^{1,853} \cdot c_s{}^{3,811} \cdot c_a{}^{-11,423}, \tag{3}$$

in this case the Gauss's criterion has the value $S_G = 0.0008970296$,

$$\Delta W_w = 0.04346 \cdot t_w{}^{-0,385} \cdot l_s{}^{0.437} \cdot c_s{}^{3,051} \cdot c_a{}^{-10.115}, \tag{4}$$

for which the Gauss's criterion has the value $S_G = 0.00000848762$, and

$$\Delta D_t = 0.142 \cdot t_w{}^{-0,242} \cdot l_s{}^{0.303} \cdot c_s{}^{-0,0933} \cdot c_a{}^{1.384}, \tag{5}$$

the Gauss's criterion has the value $S_G = 0.000007217791$.

On the base of the empirical mathematical models expressed by the Equations (3)–(5), the graphical representations from Figures 5–8 were elaborated.

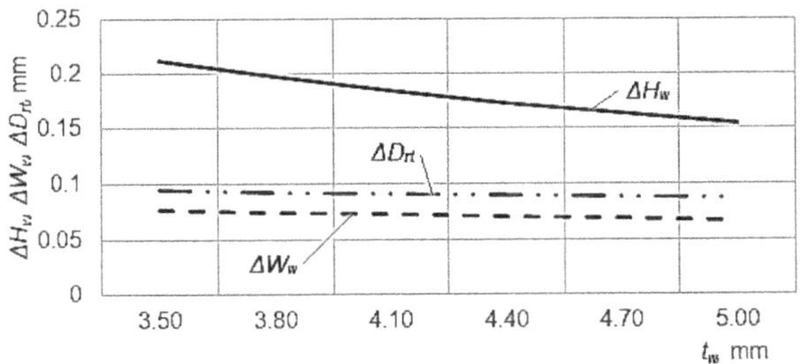


**Figure 5.** Influence exerted by the workpiece thickness $t_w$ on the deviations $\Delta H_w$, $\Delta W_w$, and $\Delta D_{rt}$ ($l_s$ = 6.5 mm, $c_s$ = 0.14 mm, $c_a$ = 0.54 mm).

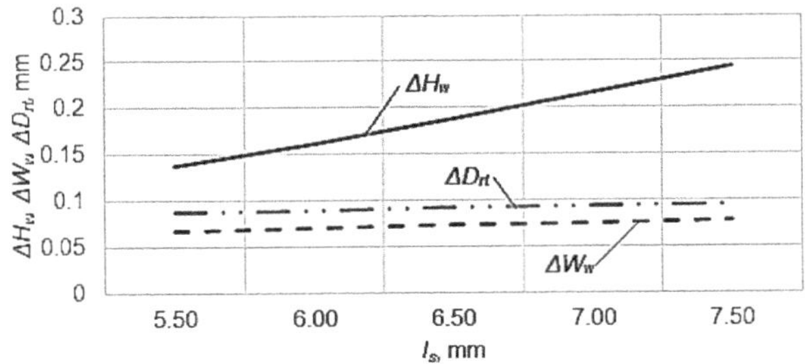


**Figure 6.** Influence exerted by the work stroke length $l_s$ on the deviations $\Delta H_w$, $\Delta W_w$, and $\Delta D_{rt}$ ($t_w$ = 4 mm, $c_s$ = 0.14 mm, $c_a$ = 0.54 mm).

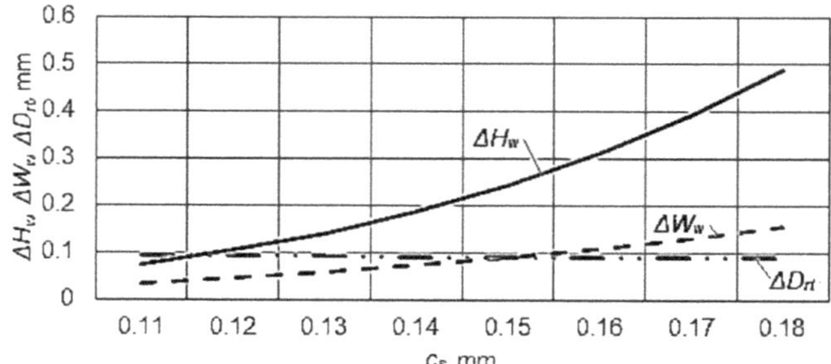

**Figure 7.** Influence exerted by the side clearance $c_s$ on the deviations $\Delta H_w$, $\Delta W_w$, and $\Delta D_{rt}$ ($t_w = 4$ mm, $l_s = 6.5$ mm, $c_a = 0.54$ mm).

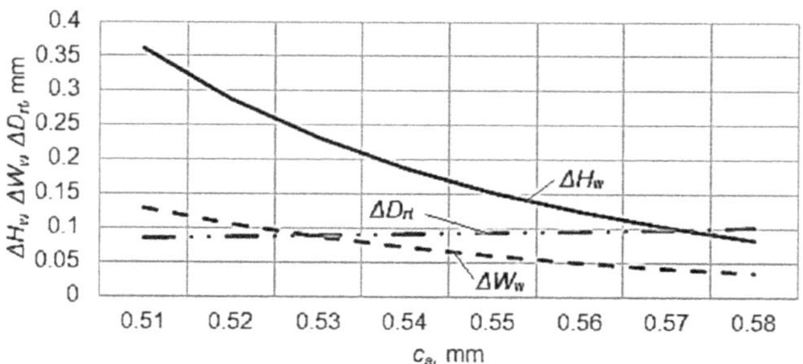

**Figure 8.** Influence exerted by the axial clearance $c_a$ on the deviations $\Delta H_w$, $\Delta W_w$, and $\Delta D_{rt}$ ($t_w = 4$ mm, $l_s = 6.5$ mm, $c_s = 0.14$ mm).

## 5. Discussion

The analysis of the empirical mathematical models (3)–(5), and of the graphical representations from Figures 5–8, facilitated the formulation of some general remarks concerning the influence exerted by the process input factors on the dimensional precision of windows cut in bearing cages.

Thus, one can notice that the increase of the workpiece thickness $t_w$ determines an increase of the machining accuracy (a diminishing of the deviations $\Delta H_w$, $\Delta W_w$, and $\Delta D_{rt}$, Figure 5), and this could be justified by the increase of the workpiece rigidity, which has as an effect a more precise cutting of the windows. In all the three mathematical empirical models, the exponents attached to the factor $t_w$ have a negative value, and the highest influence is exerted on the deviation $\Delta H_w$ of the window height $H_w$ when the exponent has a maximum value.

If the influence of the work stroke length $l_s$ is analyzed, one could notice that in all the three cases the increase of this factor value determines an increase of the deviations $\Delta H_w$, $\Delta W_w$, and $\Delta D_{rt}$ (a decrease of the machining precision, Figure 6). The fact could be justified by the longer contact between the workpiece and the cutting punch, which could contribute to a supplementary material removal from the workpiece, inclusively by developing an intense friction phenomenon. The maximum influence is exerted on the deviation $\Delta H_w$ of the window height, this meaning that the exponent attached to the factor $l_s$ has the maximum value in the mathematical empirical model that corresponds to the deviation $\Delta H_w$ of the window height.

As expected, the increase of the side clearance $c_s$ determines a decrease of the machining precision in the case of the deviations $\Delta H_w$, $\Delta W_w$ (Figure 7), when the exponents attached to the factor $c_s$ has values higher than 3. Indeed, the increase of the side clearance $c_s$ could change the behavior of the workpiece material during the cutting process, and this could determine an increase of the deviations

$\Delta H_w$ and $\Delta W_w$. In the case of the of the deviation $\Delta D_{rt}$ of the distance $D_{rt}$ between the retention thresholds, the influence is practically insignificant, with the value of the exponent attached to the factor $c_s$ being close to zero.

For relatively large values of the axial clearance $c_a$ (0.51–0.58 mm), one notices that the increase of this factor leads to a decrease of the deviations $\Delta H_w$, $\Delta W_w$ (the exponents attached to the factor $c_a$ in the empirical mathematical models having negative values) and exerts a relatively low influence on the deviation $\Delta D_{rt}$ of the distance between the retention thresholds, as can be seen in Figure 8. The strongest influence is exerted on the deviation $\Delta D_{rt}$ of the distance between the retention thresholds, the fact being highlighted both by the exponent attached to the factor $c_a$ in the empirical mathematical model of $\Delta D_{rt}$ (Equation (5)) and by the graphical representation from Figure 8.

## 6. Conclusions

The windows that exist in the bearing cage could be obtained by a forming process that involves the use of a punch and a die. Over the years, researchers have become concerned with the influence exerted by distinct factors on the machining precision that corresponds to punching processes. Only a few experimental results referred to the machining precision of the bearings cage windows. In this paper, the problem of investigating the machining precision in case of the cutting process of windows in cages for double row tapered roller bearings was addressed. The theoretical analysis highlighted that the workpiece thickness, the length of punch work stroke, and the clearance between the punch and die could be factors that are able to affect the dimensional precision of the windows cut in the bearing cage. Experimental research was developed to identify the influence exerted by the above-mentioned process input factors on the precision of some of the windows' overall dimensions and on the so-called distance between the retention thresholds. By mathematical processing of the experimental results, power type empirical mathematical relations were determined. These empirical mathematical models and the graphical representations were analyzed to obtain information concerning the influence of some process input factors on the machining precision of obtaining certain dimensions of the cage windows. Essentially, the increase of the workpiece thickness determines the increase of the machining precision, while the increase of the punch work stroke length contributes to a decrease in the machining accuracy. The increase of the side clearance leads to an increase of the cutting deviation, and the increase of the axial clearance determines the decrease of the overall dimension deviations. In the future, there is the intention to extend the experimental research so that the empirical mathematical models will be used to optimize the punching process of window cutting in bearing cages.

**Acknowledgments:** The experimental research benefited from the material support offered by the company S.C. "Rulmenți" S.A. Bârlad (România).

**Author Contributions:** M.R. conceived and materialized the experimental research, G.N. had the idea of studying the influence exerted by the experimental conditions of the dimensional accuracy of the windows achieved in bearing cages and established the experimental conditions, L.S. proposed and selected the empirical mathematical models and proposed the general structure of the paper, and O.D. achieved the mathematical processing of the experimental results and edited the paper.

**Conflicts of Interest:** The authors declare no conflict of interests.

## References

1. Kibe, Y.; Okada, Y.; Mitsui, K. Machining accuracy for shearing process of thin-sheet metals—Development of initial tool position adjustment system. *Int. J. Mach. Tools Manuf.* **2007**, *47*, 1728–1737. [CrossRef]
2. Kibe, Y.; Mitsui, K. Machining Accuracy of Punching for Thin Sheet Metal—Influence of Tool Position Misalignment and Its Detection. *J. Jpn. Soc. Precis. Eng.* **2009**, *75*, 373–378. [CrossRef]
3. Ristiawan, I.; Mahardika, M. Effect of Clearance and Punch Speed on the Cutting Surface Quality Results of a Brass Blanking on the Micropunch CNC Machine. In Proceedings of the AIP Conference Proceedings,

the 7th International Conference on Mechanical and Manufacturing Engineering, Sustainable Energy Towards Global Synergy, Jogjakarta, Indonesia, 1–3 August 2016; Volume 1831.

4.  Ardeshana, M.V.; Mehta, N.L. Design of Punch and Die for Taper Roller Bearing Cage for Multi Pocketing. *Int. J. Mech. Eng. Technol.* **2013**, *4*, 367–372.

5.  Zhao, X.; Liang, L.; Liu, Z. Tapered Roller Bearings Cage to Improvement Design. *Indones. J. Electr. Eng. Comput. Sci.* **2014**, *12*, 1855–1859. [CrossRef]

6.  English, J.A.; Maurizi, J.G. Method for Producing Windows in Cages of Roller Bearings. U.S. Patent 3,080,639 A, 1959. Available online: http://www.google.co.ug/patents/US3080639 (accessed on 29 November 2017).

7.  Ahlman, S.R. Method of Punch-Forming Windows in Bearing Retainers. U.S. Patent 3,416,211, 1962. Available online: http://www.google.co.ug/patents/US3416211 (accessed on 29 November 2017).

8.  Ichikawa, K.; Oooka, S. Needle Roller with Bearing with a Retainer and Manufacturing Method of the Retainer. Patent Application EP1741941A2, 2005. Available online: http://www.google.co.ug/patents/EP1741941A2?cl=en (accessed on 29 November 2017).

9.  Kamiji, M.; Okano, M.; Maeda, H.; Kawatake, Y.; Teshima, K.; Yabubayashi, Y.; Hosokawa, T. Roller Bearing Cage and Manufacturing Method Therefor as well as Roller Bearing Manufacturing Method. Patent Application WO2013133363 A1, 2012. Available online: https://www.google.co.ug/patents/WO2013133363A1?cl=en (accessed on 29 November 2017).

10. Mania, B.T.; Prescavage, J. Cage for a Roller Bearing and a Method of Manufacturing the Same. U.S. Patent 9,033,587 B1, 2013. Available online: https://www.google.co.ug/patents/US9033587 (accessed on 29 November 2017).

11. Gao, Y.; Villecco, F.; Li, M.; Song, W. Multi-Scale Permutation Entropy Based on Improved LMD and HMM for Rolling Bearing Diagnosis. *Entropy* **2017**, *19*, 176. [CrossRef]

12. Wu, L.; Yao, B.; Peng, Z.; Guan, Y. Fault Diagnosis of Roller Bearings Based on a Wavelet Neural Network and Manifold Learning. *Appl. Sci.* **2017**, *7*, 158. [CrossRef]

13. Li, K.; Wu, J.; Zhang, Q.; Su, L.; Chen, P. New Particle Filter Based on GA for Equipment Remaining Useful Life Prediction. *Sensors* **2017**, *17*, 696. [CrossRef] [PubMed]

14. Villecco, F.; Pellegrino, A. Entropic Measure of Epistemic Uncertainties in Multibody System Models by Axiomatic Design. *Entropy* **2017**, *19*, 291. [CrossRef]

15. Villecco, F.; Pellegrino, A. Evaluation of Uncertainties in the Design Process of Complex Mechanical Systems. *Entropy* **2017**, *19*, 475. [CrossRef]

16. Rîpanu, M.I.; Nagîţ, G.; Slătineanu, L.; Dodun, O.; Mihalache, A.M. Surface Roughness Obtained at Stamping of Bearing Cages. In Proceedings of the MATEC Web Conference on Modern Technologies in Manufacturing (MTeM 2017—AMaTUC), Cluj-Napoca, Romania, 12–13 October 2017; Volume 137. [CrossRef]

17. Rîpanu, M.I. Theoretical and Experimental Researches Regarding the Quality and Precision of Stamped Bearings Cages. Doctoral Thesis, Gheorghe Asachi Technical University, Iaşi, Romania, 2014.

18. Creţu, G. *Fundamentals of Experimental Research. Laboratory Handbook*; Gheorghe Asachi Technical University: Iaşi, Romania, 1992; pp. 27–35.

 *machines*

*Article*

# Numerical and Experimental Characterization of a Railroad Switch Machine

Dario Croccolo, Massimiliano De Agostinis, Stefano Fini *, Giorgio Olmi and Francesco Robusto

Department of Industrial Engineering, University of Bologna, Viale del Risorgimento 2, 40136 Bologna, Italy; dario.croccolo@unibo.it (D.C.); m.deagostinis@unibo.it (M.D.A.); giorgio.olmi@unibo.it (G.O.); francesco.robusto2@unibo.it (F.R.)
* Correspondence: stefano.fini@unibo.it; Tel.: +39-051-2093455

Received: 15 January 2018; Accepted: 12 February 2018; Published: 17 February 2018

**Abstract:** This contribution deals with the numerical and experimental characterization of the structural behavior of a railroad switch machine. Railroad switch machines must meet a number of safety-related conditions such as, for instance, exhibiting the appropriate resistance against any undesired movements of the points due to the extreme forces exerted by a passing train. This occurrence can produce very high stress on the components, which has to be predicted by designers. In order to assist them in the development of new machines and in defining what the critical components are, FEA models have been built and stresses have been calculated on the internal components of the switch machine. The results have been validated by means of an ad-hoc designed experimental apparatus, now installed at the facilities of the Department of Industrial Engineering of the University of Bologna. This apparatus is particularly novel and original, as no Standards are available that provide recommendations for its design, and no previous studies have dealt with the development of similar rigs. Moreover, it has wide potential applications for lab tests aimed at assessing the safety of railroad switch machines and the fulfilment of the specifications by many railway companies.

**Keywords:** railroad switch; railway junction; FEA; experimental; points

## 1. Introduction

A railroad switch machine (RSM), turnout or set of points is a mechanical installation enabling railway trains to be guided from one track to another, such as at a railway junction or where a spur or siding branches off. One of the key safety requirements of railroad switches is related to achieving a suitable resistance against any undesired movements of the points, due, for instance, to the extreme forces exerted by a passing train in the case of the needle leaned to the rail (force F in Figure 1).

Many railway companies assume a force F = 100 kN as standard. This work deals with the development of FEA models aimed at accomplishing the structural design of the RSM under the aforementioned operating load. In order to validate such models, an experimental test bench has been designed and manufactured. This comprises two ad-hoc designed fixtures that allow the accommodation of the test piece on a standard INSTRON 8500 500 kN standing press and the application of forces up to a maximum of F = 300 kN. Issues of novelty arise from the lack of studies both in the scientific and in the technical literature dealing with the development of similar fixture devices. The developed testing rig can be used not only for FEA validation purposes, but also for experimental tests aimed at warranting the safety of the RSM and the accomplishment of design requirements by most railway companies. The originality of the performed non-trivial design task arises also from the lack of specific Standards providing recommendations or reference schemes for the execution of lab tests aimed at assessing the structural response of RSM under high loads.

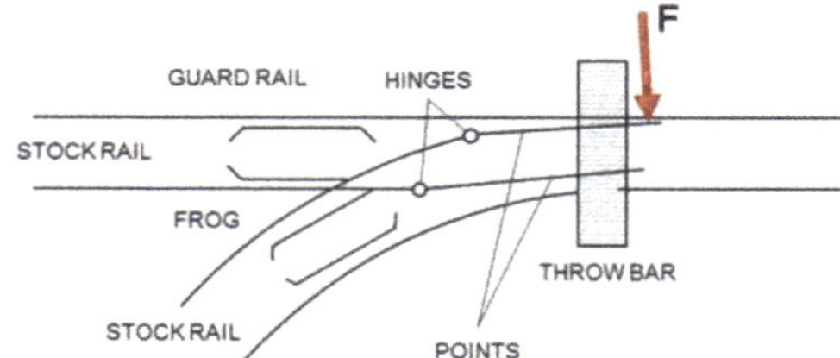


**Figure 1.** Geometry of a railroad switch.

## 2. Materials and Methods

The Alstom RSM object of the present investigation is shown in Figure 2, along with some balloons highlighting the key structural components of the machine.

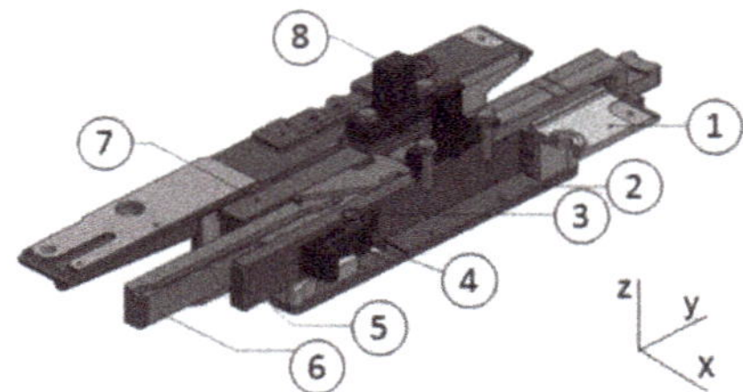


**Figure 2.** 3d model of the Alstom RSM: (**1**) body; (**2**) lower plate; (**3**) pin; (**4**) hammer; (**5**) switching rod; (**6**) cam; (**7**) detection rod; (**8**) arm.

Due to confidentiality-related issues, the working principles of the machine cannot be described in detail. The analysis was limited to the verification of the mechanism against unwanted movements of the points caused by a passing train since the system is equipped with two interlocking devices. In fact, once the full stroke has been travelled, and the points are in the open (or closed) position, the switching rod (5) is secured to the body (1) by means of a hammer (4); at the same time, the detection rod (7) is secured to the lower plate (2) by means of a slider, not represented in the picture. Therefore the locking devices come into effect preventing any movement of the rods, when an external force is applied along z-axis to the points, and thereby to the arms (8).

According to the requirements set by railway companies, the RSM should be validated under the action of a force F = 100 kN. The load application rate surely affects the response of the structure. The testing force of F = 100 kN is set by the railway company in order to account for dynamic effects. In order to attain an adequate stiffness of the test fixture, it has been dimensioned for a maximum load of 300 kN. The overall dimensions of the test piece are $900 \times 300 \times 210$ mm; therefore, the fixture was conceived in two separate parts, a lower and an upper grip, so as to achieve a certain flexibility during mounting and unmounting operations on the standing press. In order not to transmit any unwanted bending moment at the arms, the test fixture was shaped as shown in Figure 3. While the lower grip is a simple C-shaped interface between the actuator thread and the arms, the upper grip has to retain the whole RSM by means of four M20 8.8 class bolts. The bolted joint is doubly overlapped: this provision allows doubling the frictional surfaces and hence the transmissible load for a given bolt size and class [1,2]. Except for a few details, the fixture has to be arc welded, therefore a structural steel S275JR according to [3] has been chosen for its construction. All the welds were statically dimensioned according to Standard EN 1993-1-8 [4]. In order to assess the stresses and the deformations on the fixture under maximum design load (F = 300 kN), some FEA have been performed by means of the commercial code Ansys Workbench.

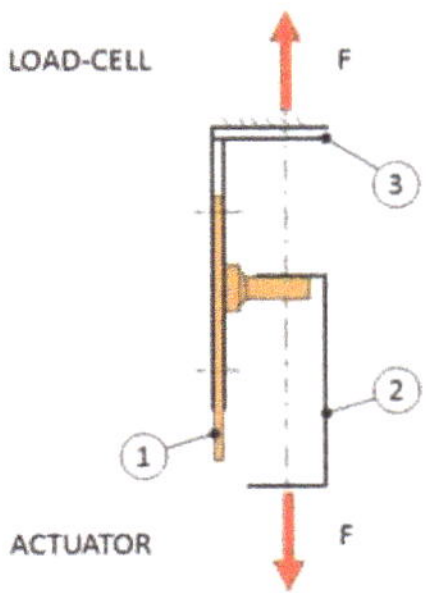



**Figure 3.** Loading scheme: (**1**) test piece (switch machine); (**2**) lower grip; (**3**) upper grip.

Figure 4a shows the boundary conditions applied to the model: the upper grip has been fixed at the upper end and loaded by two equal forces Fz = 150 kN, one at each arm. The model has been meshed with SOLID187 Tetrahedral and Hexahedral elements, Figure 4b. The material is a structural steel, whereas the bonded contacts are managed by means of the pure penalty contact algorithm, with the normal stiffness factor set to FKN = 0.01, following the lines suggested by [5,6]. The analysis and model parameters are summarized in Table 1.

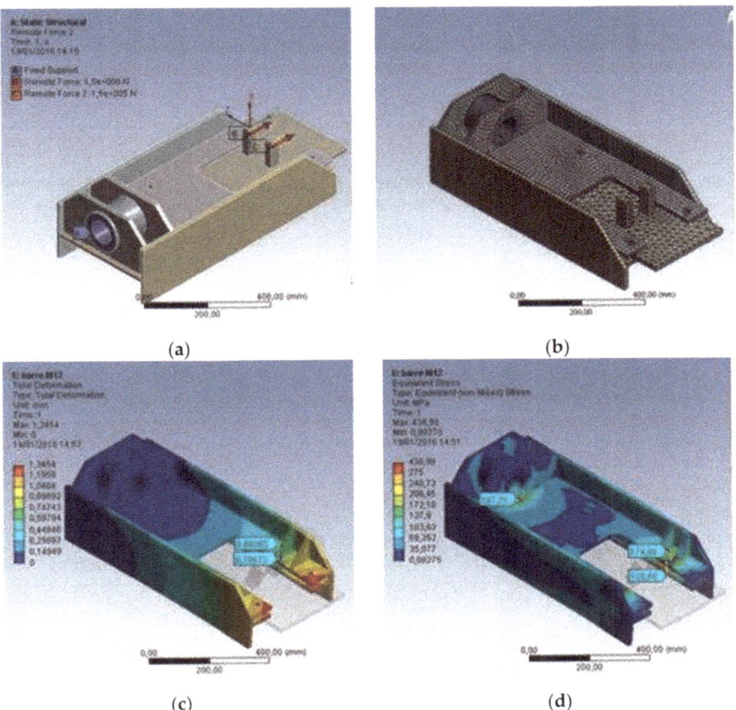

(**a**)

(**b**)

(**c**)

(**d**)

**Figure 4.** FEA on the fixture upper grip: (**a**) Boundary conditions; (**b**) mesh; (**c**) total deformation; (**d**) equivalent von Mises stress.

Table 1. Analysis and model parameters.

| | Number of Nodes | Element Types | Elastic Modulus | Poisson Ratio |
|---|---|---|---|---|
| | [-] | [-] | E [Mpa] | v [-] |
| Fixture | 80,000 | SOLID187 (Tetrahedral and Hexahedral) | 200,000 | 0.3 |

As can be appreciated by looking at Figure 4c, the total deformation is Δtot_max = 1.3 mm, whereas the maximum von Mises equivalent stress remains below 190 MPa (see Figure 4d); such a stress level is well below the material yield point SY = 275 MPa. Since the model is linear, a maximum deformation of about Δtot_nom = 0.4 mm can be expected at nominal load, which is deemed acceptable.

The assembly procedure of the test rig requires quite a number of subsequent operations, briefly summarized in Figure 5. In particular, Figure 5d shows a detailed view of the arms of the machine when these are clamped by the lower grip. When the assembly is done, the load cell undergoes zero calibration and the test can begin. The main goal of the experiment is to provide a validation of the FEA models of the RSM that will be described in the following. A secondary aim of the experimentation is to determine how much of the total load is borne by the switching rod and how much by the detection rod. In order to accomplish this twofold task, three components of the RSM were instrumented by strain gauges: the two arms and the pin (see Figure 6).

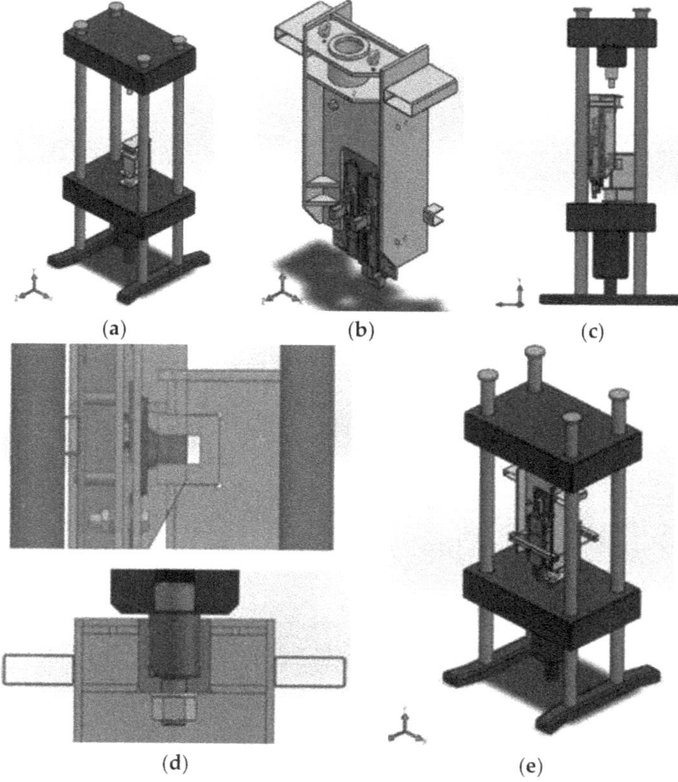

(a)       (b)       (c)

(d)       (e)

**Figure 5.** Test arrangement: (**a**) Placement of the lower grip, (**b**) pre-mounting of the upper grip with the test piece; (**c**) placement of the upper grip with a forklift; (**d**) details of the assembly; (**e**) final configuration.

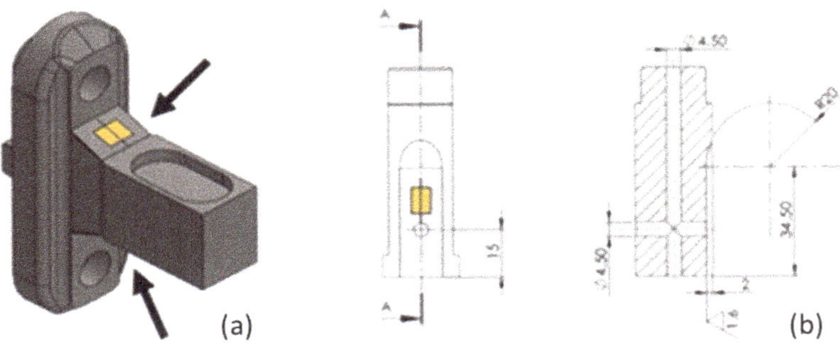


**Figure 6.** (**a**) Placement of the strain gauges on the arm and (**b**) on the reworked pin.

The pin and arms had been previously reworked in order to accommodate the sensors (Vishay Precision Group J2A-XXS047K-350); in particular, the pin required both milling and boring operations in order to achieve a plane surface for the application of the strain gauge, as well as a passage for the cables. The arms were instrumented by means of two strain gauges each; the strain gauges were connected in a half-bridge fashion to the Wheatstone circuit. The pin was instrumented by means of a single strain gauge; a dummy gauge, which served as a temperature drift compensator, was glued to an identical, unloaded pin placed in the testing room. The adhesive used for the installation was the M-BOND 200 by Vishay Precision Group. All the sensors were installed by a certified operator, following the guidelines suggested by the Standards [7–9]. Data acquisition was managed by means of the NI 9237 sampling card plugged into a NI cDAQ-9184 carrier. The FEA model of the RSM was developed by means of the Ansys code V.17. Due to the complexity of the assembly, submodeling was leveraged, by considering half a model at a time, as if the machine were cut along its mid-plane, normal to the $x$-axis. In this way, it was possible to find a satisfactory balance between accuracy and computational cost. Both models were meshed with tetrahedral elements SOLID187, switching rod side (Figure 7), and detection rod side (Figure 8). The analysis and model parameters are summarized in Table 2.

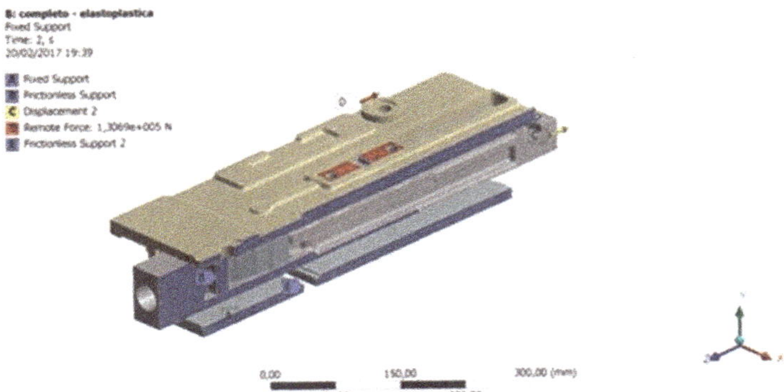


**Figure 7.** Boundary conditions for the half model comprising the switching rod.

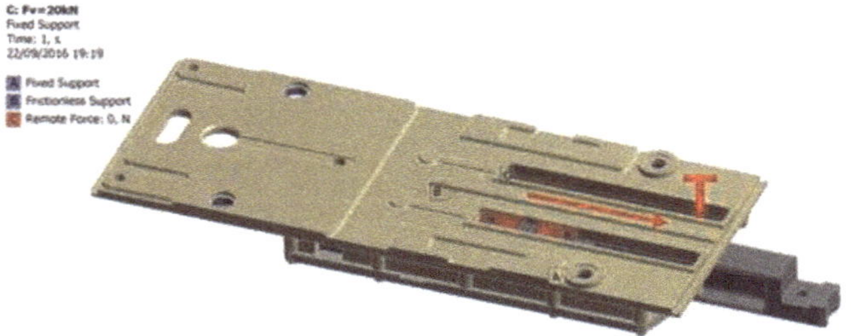



**Figure 8.** Boundary conditions for the half model comprising the detection rod.

**Table 2.** Analysis and model parameters.

| | Number of Nodes | Element Types | Steel Components | | Cast Iron Components | |
|---|---|---|---|---|---|---|
| | | | Elastic Modulus | Poison Ratio | Elastic Modulus | Poisson Ratio |
| | [-] | [-] | E [Mpa] | $\nu$ [-] | E [Mpa] | $\nu$ [-] |
| RSM switching rod side | 360,000 | SOLID187 (Tetrahedral and Hexahedral) | 200,000 | 0.3 | 169,000 | 0.275 |
| RSM detection rod side | 485,000 | SOLID187 (Tetrahedral and Hexahedral) | 200,000 | 0.3 | 169,000 | 0.275 |

In the case of the switching rod, the stresses on the pin and those on the hammer were sampled, and subsequently compared with the experimental outcomes. In the case of the detection rod, the stresses on the pins that connect the lower plate to the body were examined, as a function of the actual bolt preload of the joint.

## 3. Results and Discussion

The results from a tensile test carried out on the ad-hoc developed test bench are shown in Figure 9.

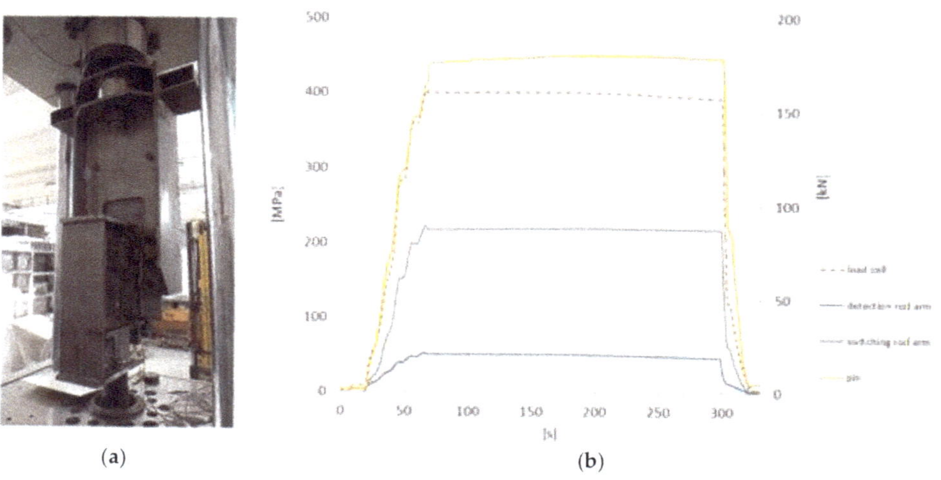



(a)　　　　　(b)

**Figure 9.** (**a**) View of the test bench and (**b**) plot of the results in terms of stresses on the pin and on the arms and force at the load cell.

Figure 9 reports the data relevant to a test run until a maximum force of F = 160 kN. At the peak load, one of the pins connecting the lower plate with the body failed. The first outcome of the experiment is the knowledge of the force distribution on the two arms: the great majority of the total force (82%) reaches the body by passing through the chain of components named the switching rod, the pin and the hammer. The remaining part (18%) passes through the detection rod, the slider and the lower plate, eventually reaching the body. Running each of the FEA models by applying the appropriate fraction of the total load to the arm under investigation, it was possible to validate the numerical results. For example, looking at Figure 10a, it is possible to observe the equivalent stresses calculated according to the von Mises criterion on the half machine comprising the switching rod. Figure 10b reports the σY bending stresses on the pin that supports the hammer, when this sub-system is loaded with 82% the total load F = 160 kN.

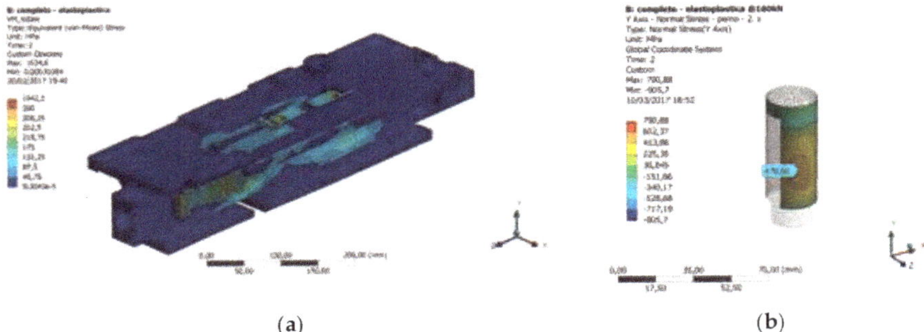

(a)  (b)

**Figure 10.** (**a**) von Mises stress plot on the half machine—switching rod side, and (**b**) bending σY stresses on the pin.

As can be appreciated from Figure 10b, the numerical peak of the bending stress on the pin (σY_FEA = 477 MPa) is very close to that measured during the experimental test on the same component (σY_EXP = 450 MPa, see Figure 9). The error, calculated according to Equation (1), is acceptable.

$$e\% = \frac{\sigma_{Y\_FEA} - \sigma_{Y\_EXP}}{\sigma_{Y\_EXP}} \cdot 100 = 6\% \tag{1}$$

Once the FEA model has been validated, it can be used for carrying out some comparisons considering, for example, the joint between the lower plate and the body. Such joints comprise a pattern of eight M8 8.8. screws, working in parallel with a couple of parallel pins of d = 6 mm diameter, manufactured according to Standard [10]. It can be assumed that this joint must withstand the shearing load transmitted by the slider to the body via the lower plate. These pins are coupled with interference (H7/m6). Since the screws are tightened under preload control upon assembly, and some uncertainties with regard to the friction coefficients cannot be avoided [11,12], the load borne by the parallel pins may vary depending on the effective preload of the screws and on the friction coefficient at the interface between the body and the plate. In order to estimate such variation, some parametric analyses were run, for example by imposing different preload levels on the screws. The screw preload was assigned via the preload tool available in the Ansys Workbench environment. Figure 11 reports a plot of the amount of shearing force borne by the switching rod side pin (T′swi) and by the detection rod side pin (T′det) as functions of the actual screw preload Fv. Each of the dashed lines represents the force acting on the relevant parallel pin, whereas the solid lines represent the fraction of load borne by the generic pin.

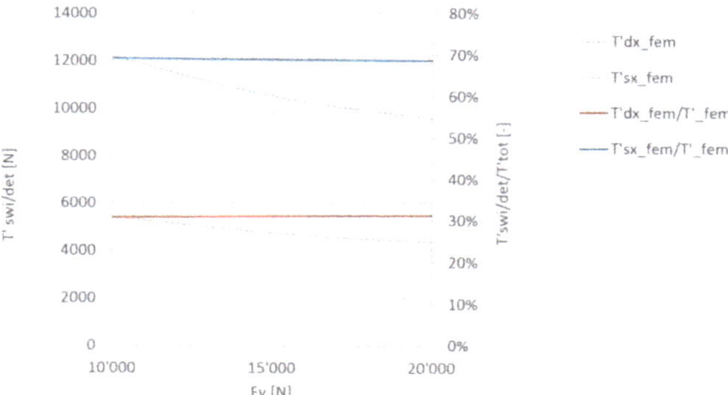

**Figure 11.** Shearing force on the switching/detection rod side parallel pin versus screw preload.

It can be seen that the most loaded pin is that on the detection rod side (closer to the slider), regardless of the screw preload. Nonetheless, the magnitude of the load borne by the pins decreases as the screw preload increases: a preload limit of Fv = 20 kN is assumed based on the provisions of Standard [13] for M8, 8.8 class screws. Based on different preload levels, it is also possible to extract a plot of the von Mises stresses on the most loaded parallel pin, as shown in Figure 12.

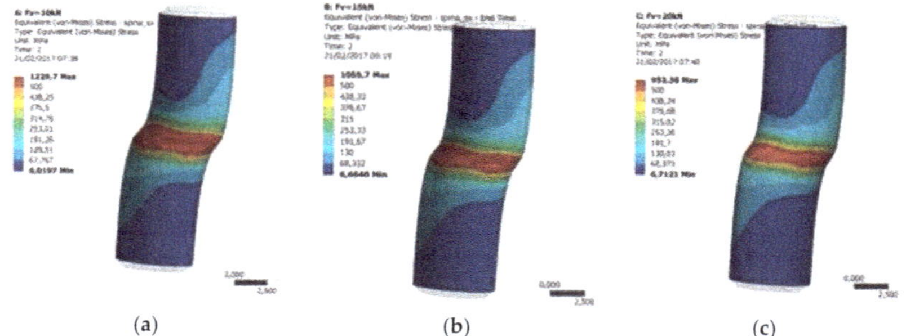

**Figure 12.** von Mises stress plot on the detection rod side parallel pin at a screw preload of (**a**) 10 kN; (**b**) 15 kN; (**c**) 20 kN.

The equivalent stress calculated by FEA on the most loaded parallel pin is compatible with the failure event, which took place during the experiment at a total load of F = 160 kN. The strength of the pins could be modified by changing the coupling system, increasing the interference or adopting a different coupling technique. Based on the literature, a valid alternative could be making use of anaerobic or epoxy adhesives, which would make it possible to significantly increase the actual mating area with a positive outcome in terms of the overall strength. This point has been tackled experimentally in papers [14–16], which also provide tips regarding the proportioning of the joint upon its design.

## 4. Conclusions

From a designer's standpoint, the present work achieved a twofold result: (i) an experimental setup has been designed, manufactured and calibrated, which is novel and original and will be useful for subsequent experimentations on other products of the same family; (ii) numerical tools have been developed and validated, with respect to experimental data. These models allow the designer to evaluate the effect of structural changes early, hence reducing the time to market of new machines.

**Acknowledgments:** The authors gratefully acknowledge Eng. Marcello Andrenacci, Eng. Leonardo Bozzoli, Eng. Francesco Muscatello and Eng. Francesca Sopranzetti at Alstom Ferroviaria SpA for having made this research possible. The authors would also like to acknowledge Eng. Francesco Vai, laboratory director at the Department of Industrial Engineering, University of Bologna, for his fundamental contribution to the experimental activities.

**Author Contributions:** D.C. and M.D.A. conceived and designed the experiments; S.F. and M.D.A. performed the experiments; M.D.A., S.F., and F.R. performed the numerical analyses; G.O. analyzed the data; S.F., and M.D.A. provided reagents, materials and analysis tools; G.O. and M.D.A. wrote the paper.

**Conflicts of Interest:** The authors declare no conflict of interest.

## References

1. Niemann, G.; Winter, H.; Hohn, B.R. *Maschinenelemente: Band 1: Konstruktion und Berechnung von Verbindungen, Lagern, Wellen*; Springer-Verlag: Berlin, Germany, 2005.

2. De Agostinis, M.; Fini, S.; Olmi, G. The influence of lubrication on the frictional characteristics of threaded joints for planetary gearboxes. *Proc. Inst. Mech. Eng. Part C J. Mech. Eng. Sci.* **2016**, *230*, 2553–2563. [CrossRef]

3. *Hot Rolled Products of Structural Steels, Part 2: Technical Delivery Conditions for Non-Alloy Structural Steels*; UNI EN 10025-2:2005; British Standards Institution: London, UK, 2005.

4. *Eurocode 3, Design of Steel Structures, Part 1-8: Design of Joints*; UNI EN 1993-1-8:2005; Europian Comittee for Standardisation: Brussels, Belgium, 2005.

5. Croccolo, D.; De Agostinis, M.; Vincenzi, N. Structural analysis of an articulated urban bus chassis via FEM: a methodology applied to a case study. *Stroj. Vestnik J. Mech. Eng.* **2011**, *57*, 799–809. [CrossRef]

6. Zhao, X.; Li, F.; Fan, Y.; Liu, Y. Fatigue behavior of a box-type welded structure of hydraulic support used in coal mine. *Materials* **2015**, *8*, 6609–6622. [CrossRef] [PubMed]

7. *Non Destructive Testing—Inspection by Strain Gauges: Terms and Definitions*; UNI EN 10478-1:1996; Italian Standards: Milano, Italy, 1996.

8. *Non Destructive Testing—Inspection by Strain Gauges: Selection of Strain Gauges and Accessory Components*; UNI EN 10478-2:1998; Italian Standards: Milano, Italy, 1996.

9. *Non Destructive Testing—Inspection by Strain Gauges: Strain Gauge Installation and Checking*; UNI EN 10478-3:1998; Italian Standards: Milano, Italy, 1998.

10. *Parallel Pins, of Hardened Steel and Martensitic Stainless Steel*; ISO 8734; International Organization for Standardization: Geneva, Switzerland, 1997. (Dowel pins)

11. Croccolo, D.; De Agostinis, M.; Fini, S.; Olmi, G. Tribological properties of bolts depending on different screw coatings and lubrications: An experimental study. *Tribol. Int.* **2017**, *107*, 199–205. [CrossRef]

12. Eccles, W.; Sherrington, I.; Arnell, R.D. Frictional changes during repeated tightening of zinc plated threaded fasteners. *Tribol. Int.* **2010**, *43*, 700–707. [CrossRef]

13. *Mechanical Properties of Fasteners Made of Carbon Steel and Alloy Steel—Part 1: Bolts, Screws and Studs with Specified Property Classes—Coarse Thread and Fine Pitch Thread*; ISO 898-1; International Organization for Standardization: Geneva, Switzerland, 2009.

14. Croccolo, D.; De Agostinis, M.; Mauri, P.; Olmi, G. Influence of the engagement ratio on the joint strength of press fitted and adhesively bonded specimens. *Int. J. Adhes. Adhes.* **2014**, *53*, 80–88. [CrossRef]

15. Croccolo, D.; De Agostinis, M.; Fini, S.; Olmi, G. Influence of the engagement ratio on the shear strength of an epoxy adhesive by push-out tests on pin-and-collar joints: Part I: Campaign at room temperature. *Int. J. Adhes. Adhes.* **2016**, *67*, 69–75.

16. Croccolo, D.; De Agostinis, M.; Fini, S.; Olmi, G. Influence of the engagement ratio on the shear strength of an epoxy adhesive by push-out tests on pin-and-collar joints: Part II: Campaign at different temperature levels. *Int. J. Adhes. Adhes.* **2016**, *67*, 76–85. [CrossRef]

*Article*

# A Methodology for the Lightweight Design of Modern Transfer Machine Tools

**Dario Croccolo [1], Omar Cavalli [1], Massimiliano De Agostinis [1], Stefano Fini [1,*], Giorgio Olmi [1], Francesco Robusto [1] and Nicolò Vincenzi [2]**

[1] Department of Industrial Engineering, University of Bologna, Viale del Risorgimento 2, 40136 Bologna, Italy; dario.croccolo@unibo.it (D.C.); omar.cavalli2@unibo.it (O.C.); m.deagostinis@unibo.it (M.D.A.); giorgio.olmi@unibo.it (G.O.); francesco.robusto2@unibo.it (F.R.)

[2] Giuliani a Bucci Automations S.p.A. Division, Via Granarolo 167, 48018 Faenza RA, Italy; nicolo.vincenzi@igmi.it

* Correspondence: stefano.fini@unibo.it; Tel.: +39-051-209-3455

Received: 10 December 2017; Accepted: 11 January 2018; Published: 14 January 2018

**Abstract:** This paper deals with a modern design approach via finite elements in the definition of the main structural elements (rotary table and working unit) of an innovative family of transfer machine tools. Using the concepts of green design and manufacture, as well as sustainable development thinking, the paper highlights the advantages derived from their application in this specific field (i.e., the clever use of lightweight materials to allow ruling out high-consumption hydraulic pump systems). The design is conceived in a modular way, so that the final solution can cover transfers from four to 15 working stations. Two versions of the machines are examined. The first one has a rotary table with nine divisions, which can be considered as a prototype: this machine has been studied in order to set up the numerical predictive model, then validated by experimental tests. The second one, equipped with a rotary table with 15 divisions, is the biggest of the range: this machine has been entirely designed with the aid of the previously developed numerical model. The loading input forces for the analyses have been evaluated experimentally via drilling operations carried out on a three-axis CNC unit. The definition of the design force made it possible to accurately assess both the rotary table and the working units installed in the machine.

**Keywords:** transfer machine; rotary table; working unit; green design; finite element; machine tool; minimal quantity lubrication

---

## 1. Introduction: Green Design Applied to Machine Tools

In order to be competitive in today's business world, more and more companies have to plan their activities by thinking about energy consumption and resource saving. The themes of green design are compelling in the modern design of machine tools. In practical aspects, up to 10 years ago the goal of machine tool manufacturers was to improve the performance of machine tools in terms of availability, reliability, dimensional accuracy, and precision. To achieve these targets, machine tools have become increasingly complex and automated in their design. These changes resulted in increasing energy requirements, which lead to rising power costs and limited access to resources (particularly fossil fuels), and run counter to increasing environmental consciousness among customers and stricter government regulations. For instance, specifications for the purchase of machine tools set out by automotive companies have dedicated chapters dealing with energy consumption and the correct design of electrical motors, compressed air circuit, hydraulic pumps, etc., in order to optimize the energy-related costs and the environmental impact of the production. Under the label of "energy management", the following points are usually found:

1. No compressed-air motors should be used;

2. It is not allowed to use compressed air for blowing functions, for cleaning or cooling;
3. Drive systems for pumping liquids, such as hydraulic pumps shall have the option to control the input power according to the working requirements;
4. Pumps shall fulfill the criteria according to the EC regulation no. 641/2009 [1];
5. The power rating of the electric motors is to be adjusted to the mechanical power requirement of the machine. If a bigger motor has to be selected because of a performance grading, oversizing cannot overcome 30%;
6. The characteristics and design of the motors must be adjusted to the working conditions. The motors must be designed for an on-period covering the entire mission time. The performance shall be measured so that the motor is used for up to maximum 85% of its performance range;
7. The supplier shall indicate and offer potential optimization strategies for energy saving and recovery.

Machine tool energy consumption may be reduced in any of the four areas of its life cycle: (i) manufacturing (design and production of the parts), (ii) transportation (design for assembly/disassembly to allow standard transport), (iii) use (energy used to produce parts), or (iv) end of life (recycled materials). The most important phases are (i) and (iii): design-level changes are able to provide the greatest flexibility and therefore potentially offer the best opportunities for energy savings. The modern approach to the metal cutting operations such as the use of Minimal Quantity Lubrication (MQL) or the dry cutting [2–12] are able to dramatically cut consumption by nearly eliminating the impact of cutting fluids. Just for reference, it is possible to compare the need for a $\varnothing$ = 12 mm standard drill in the presence of emulsion at $p$ = 30 bar coolant pressure and for a MQL drill: 900 L/h versus 15 mL/h. Furthermore, the presence of the emulsion needs a dedicated system for the filtration of the fluid from the chips and of volumetric (high consumption) pumps to supply the high-pressure internal coolant. A breakdown of the consumption, split between the different power units in a machining center, is shown in Figure 1.

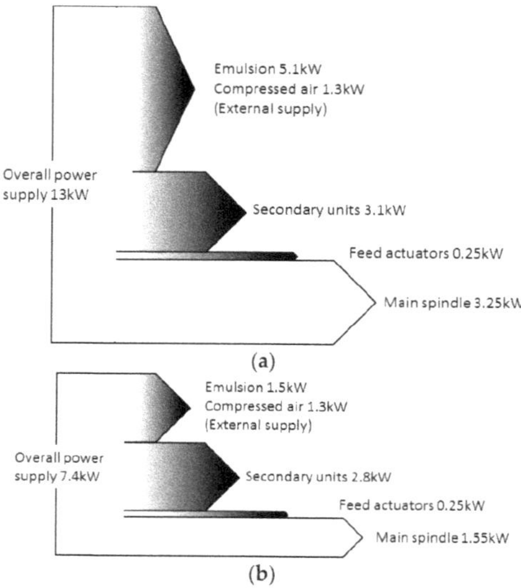


**Figure 1.** Breakdown of the average power supply needed for the production of a reference component: (**a**) roughing; (**b**) finishing. Gray shaded arrows represent energy consumption shares other than the main spindle of the machine.

Based on this scenario, a new line of drilling/milling machines has been conceived in order to fulfill the aforementioned requirements. Particularly, the main topics of the new solution can be summarized as follows:

- High dynamic performances: the idea is to build the main structural parts from lightweight materials, in order to maximize the ratio between the "chip to chip time" and the servo-motor consumption. An accurate design of the components must be carried out, in order to warrant adequate stiffness of the structure, which in turn leads to dimensional accuracy of the workpiece;
- Accurate selection of the CNC motors for the movements of the axis (as a rule of thumb, a maximum motor oversize of 30% is allowed in modern design);
- Elimination of the high-consumption hydraulic system.

A new family of transfer machine tools has been designed by Giuliani, based on the guidelines mentioned above. A high performance torque motor is mounted on each transfer machine: it makes it possible to reach high values of torque with limited energy consumption. Every component involved in the motion of the transfer is designed to be as light as possible, according to the required stiffness, to preserve the required tolerances on the finished parts. Each transfer machine consists of a number of machining stations being equipped with a three-axis reconfigurable tooling unit. The three-axis unit geometry has been recently redesigned, in order to improve its performance: this optimization is based on the reduction of the carriage weight, achieved using aluminum alloys, rather than cast iron for these parts, while keeping the stiffness of the parts unchanged. The MQL machining approach has been followed as well, to achieve both energy and coolant saving.

Leveraging the concepts of green design and manufacture, as well as sustainable development thinking, the paper highlights the advantages derived from their application in this specific field: Two versions of the machines are examined. The first one, with nine divisions, which can be considered as a prototype: this machine has been studied in order to set up the numerical predictive model, then validated by experimental tests. The second one, equipped with a bigger rotary table with 15 divisions, has been entirely designed with the aid of the previously developed numerical model. To the best of the authors' knowledge, limited to the case of transfer machine tools, no works based on the same approach are currently available in the literature.

## 2. Setup of the FE Models for the Transfer Machine with Nine Divisions

### 2.1. Experimental Determination of the Operating Loads

The three-axis unit has been designed, so that it is able to efficiently withstand a spindle thrust force in the order of $F = 1$ kN. This reference value has been chosen, following a preliminary experimental campaign, aimed at the estimation of the maximum thrust forces during the most frequent as well as the most demanding machining operations on the workpiece. In particular, a highly critical task, in terms of the load acting on the transfer, is drilling a hole under an MQL strategy. The outcome of the experiment was that 1 kN could be regarded as a proper threshold, considering for instance the drilling of a 7-mm hole on a round bar made of 16MnCrS5 steel or of a 12-mm hole on a CuZn39Pb3 brass component. In all the studied cases, which can be considered within the conventional applications of the transfer machine, the most recommended values of feed rate have been selected, also based on [13]. The estimated reference force $F$ is going to be assumed in the following numerical analyses, in order to calculate the displacements of the transfer machine under load application.

### 2.2. Experimental Characterization of the Stiffness of the Reference Transfer

The transfers involved in the investigation consist of two tables (Figure 2), whose outer diameter is $D_{out} = 1680$ mm. The lower one spins around the $y$-axis, actuated by the torque motor; whereas the upper one is fixed to the frame and supports the three-axis units. In the following, they will be referred to as the "rotating table" and "fixed table," Respectively. Supports are bolted under the rotating table in a number equal to the workstations. Each support carries a pneumatic vise that enables to safely grip the workpiece. The mechanical group described above is fixed to the ground at the base of the column.

**Figure 2.** Fixed and rotary tables of the Transfer Machine with nine divisions.

Some experimental measurements have been carried out, in order to evaluate the vertical displacements of the supports of the Transfer Machine with nine divisions, when a vertical load is applied on a support along $y$-axis: the measured displacements would then be useful for the validation of the numerical model. Three dial gauges with a resolution of 0.01 mm have been placed at the positions shown in Figure 3.

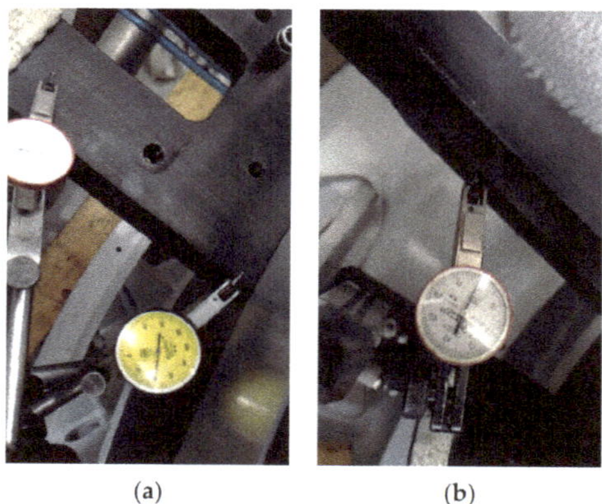

(a)                    (b)

**Figure 3.** (**a**) Dial gauges at positions 1 and 3 and (**b**) at position 2.

Then, a mass $m$ = 50 kg has been applied on the support object of investigation. The position of each dial gauge and the position of the mass are shown in Figure 4. The experimental data are shown in Table 1.

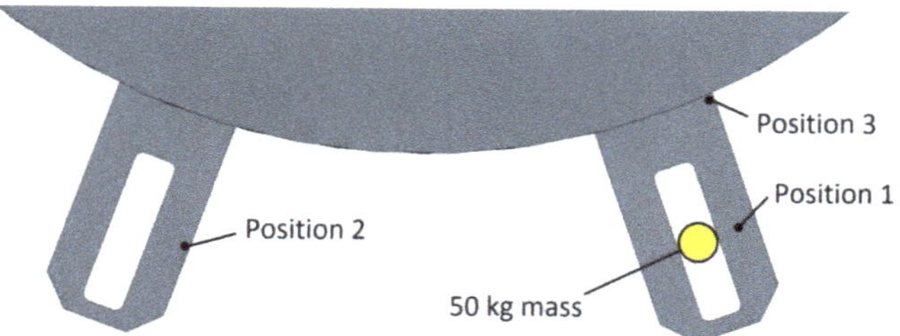


**Figure 4.** Positions of the dial gauges and of the applied mass.

**Table 1.** Dial gauge readings.

| Position | $\Delta y$ (mm) |
|:---:|:---:|
| 1 | −0.06 |
| 2 | −0.02 |
| 3 | −0.04 |

## 2.3. Tuning of the Stiffness Parameter of the FE Model

At first, the CAD geometry of the Transfer Machine with nine divisions has been simplified, in order to reduce the computational effort without significantly affecting the accuracy of the results: it must be remarked that the accurate evaluation of the stresses at notches or joints is beyond the scope of the present analysis. In fact, previous investigations carried out on similar machines showed that no significant stresses are generated on the key components of the frame during operation. Hence, the present analysis focuses just on the stiffness performance of the structure. Due to geometrical and loading symmetry conditions, a half of the geometry has been considered. A frictionless support has been applied on the symmetry plane, as shown in Figure 5a, in order to enforce the symmetry condition. Moreover, the base of the column has been constrained by means of a fixed support (Figure 5b): in the actual application, such a surface is constrained to the lower part of the frame, which can be considered perfectly rigid.

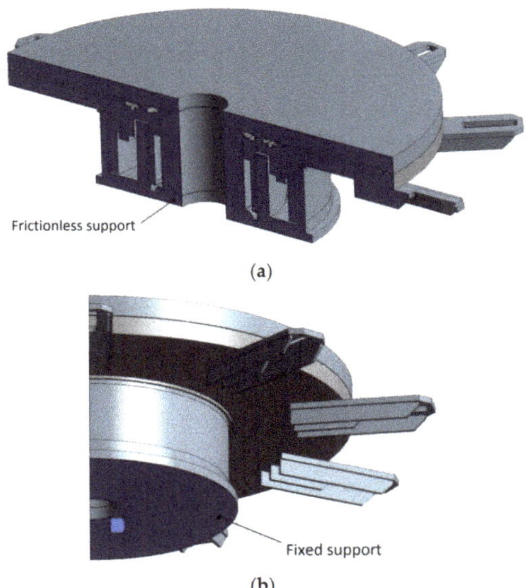


**Figure 5.** (a) Frictionless support on the symmetry plane; (b) fixed support at the base of the column.

The materials assigned to the parts are AlMg0.7Si ($E$ = 69.5 GPa, $v$ = 0.33) for the rotating table and a structural steel ($E$ = 200 GPa, $v$ = 0.30) for any other part. The geometry has been meshed with SOLID187 tetrahedral and hexahedral elements, for a total node count $n \approx 95{,}000$. All the contacts in the model are set as bonded, assuming a pure penalty formulation and a normal stiffness factor controlled by the software (FKN = 1, [14,15]). The applied force $F_y$ is calculated according to Equation (1), where term 2 in the denominator is due to the symmetry of the model.

$$F_y = \frac{9.81 \times 50}{2} = 245.25 \text{ N} \tag{1}$$

$F_y$ acts on the upper surface of the reference support, as shown in Figure 6.

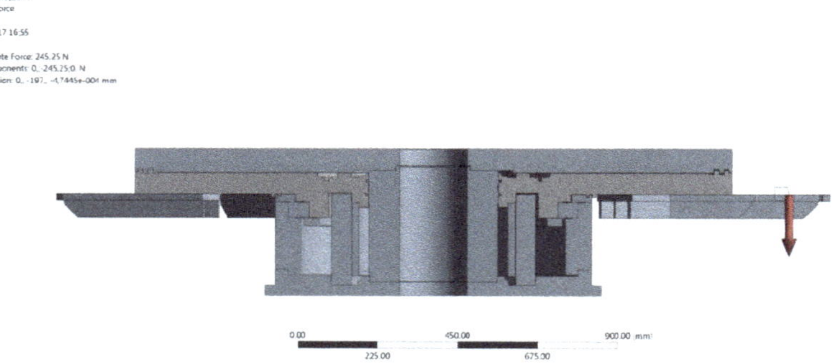

**Figure 6.** Application of $F_y$ to the reference support.

The displacements yielded by the numerical simulation and measured at the same positions of the dial gauges, are shown in Figure 7 and summarized in Table 2, along with the percentage error with respect to the experimental ones.

The results reveal that the numerical model, defined according to the basic settings reported above, is stiffer than the actual machine: hence, it has been decided to improve the model, by tuning the key parameters that affect its stiffness. The rotating table is supported by a double-row angular contact roller bearing. In the previous analysis, the bearing has been modeled as a unique ring made of steel: this approximation leads to overestimating its stiffness. Therefore, it has been decided to model the bearing by means of a single ring made of an elastic, isotropic material, whose Young's modulus has to be determined upfront by the following steps. First, (i) a FE analysis of the ring alone is prepared, by assuming its elastic modulus as a parameter and the ring is constrained, replicating the actual application and loaded by an axial thrust; (ii) the displacement of the force application surface is recorded; (iii) the experimental displacement provided by the bearing manufacturer is read-in. The manufacturer usually provides plots like that shown in Figure 8, where several curves express the axial displacement of the bearing as a function of the applied thrust load. Each curve is relevant to a value of the assembly preload $\delta$ of the bearing: for the present application, $\delta$ = 15 μm; (iv) the Young's modulus of the ring is adjusted until the FE calculated displacement matches the experimental axial displacement for given axial thrust.

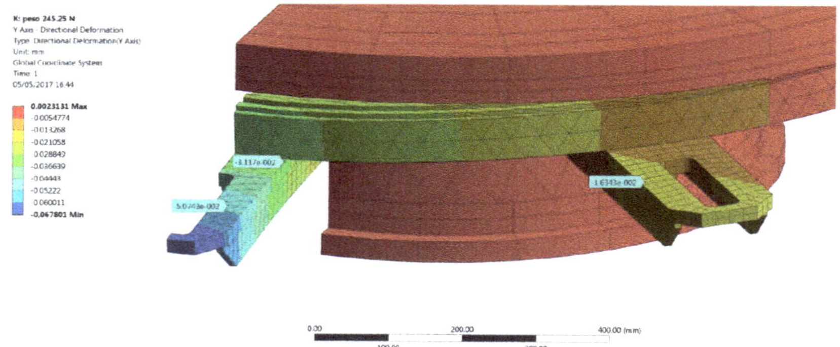

**Figure 7.** FE displacements (scale factor ×1000).

**Table 2.** FEM displacements and percentage errors with respect to the experimental data.

| Position | $\Delta y_{FEM}$ (mm) | $\Delta y_{exp}$ (mm) | Error (-) |
|----------|----------------------|----------------------|-----------|
| 1 | −0.051 | −0.060 | −15% |
| 2 | −0.016 | −0.020 | −20% |
| 3 | −0.031 | −0.040 | −22.5% |

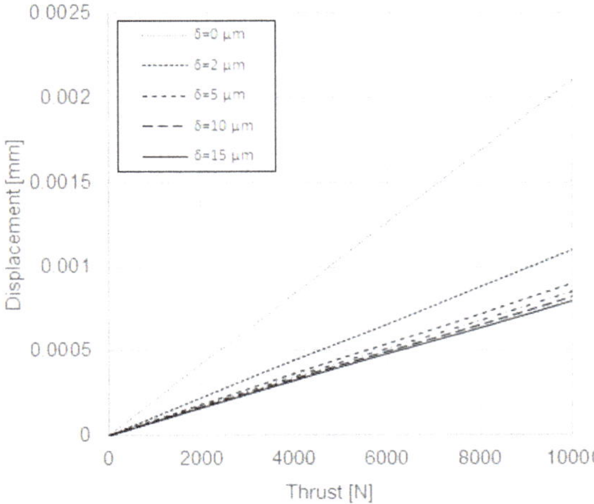


**Figure 8.** Thrust–displacement diagram of the double-row angular contact roller bearing, supplied by the manufacturer.

The static structural analysis has been carried out on a half-ring geometry, as shown in Figure 9a: the ring has been constrained with a fixed support on its lower external surface Figure 9b; the upper inner surface of the ring has been loaded with a 10 kN load (Figure 9c).

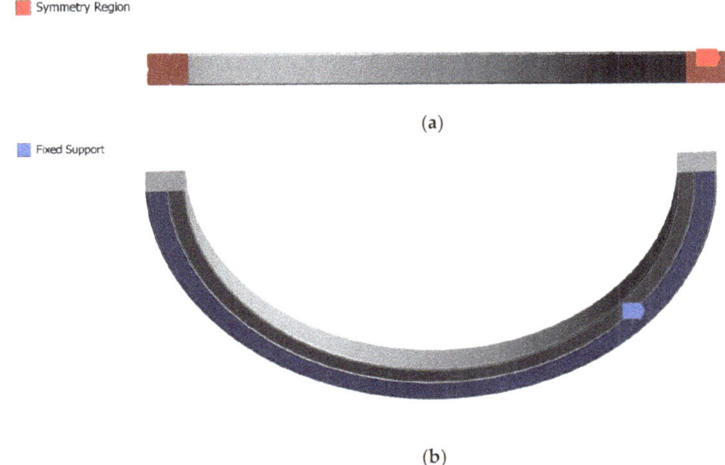


**Figure 9.** *Cont.*

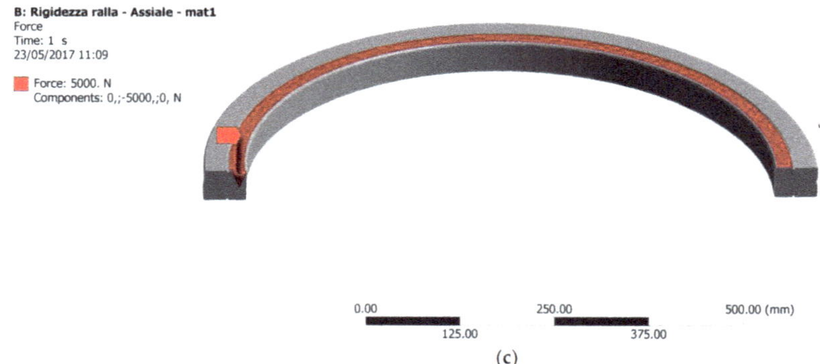


(c)

**Figure 9.** (**a**) Symmetry region, (**b**) fixed support, (**c**) load application surface.

As explained above, a set of FE analyses have been run varying the input parameter (equivalent Young's modulus of the ring) until the displacement of the loaded surface equaled $8.5 \times 10^{-4}$ mm, which is the value reported in Figure 8 for a thrust load of 10 kN and a preload of 15 µm. Such a displacement value was reached by setting an equivalent Young's modulus of $E_{ring}$ = 39 GPa. Then, a finite element analysis (FEA) of the whole Transfer Machine with nine divisions, including the ring with the above specified $E_{ring}$, was run. The displacements at the reference points (Figure 10) are substantially unvaried with respect to the previous simulation, with differences smaller than 1 µm in terms of displacement. Therefore, another modification had to be made to the numerical model in order to make it more consistent to the actual machine stiffness.

Since a correct estimate of the stiffness of a complex assembly is often related to an accurate modeling of its mechanical joints [16–18], the contact conditions between the support and the rotating table have been modified in order to achieve a more accurate FE model. Hence, the bonded contact between the rotary table and the support has been replaced by a set of four M12 12.9 class screws (Figure 11): this bolt pattern replicates that being actually used on the Transfer Machine with nine divisions. The screws have been modeled as solids, and the axial preload has been imposed by the bolt preload tool available in the Ansys Workbench environment.

The contact at the interface between the support and the rotating table, as well as the one between the underhead of the screws and the support, has been set as frictional, with a friction coefficient $\mu$ = 0.2 [19]. The contact formulation has been set as pure penalty with a normal stiffness factor controlled by the software (FKN = 1). The terminal portion of the shank and the "threaded" hole on the rotating table have been joined by means of a bonded contact. The analysis has been divided into two steps: at the first step, a preload of $F_i$ = 70 kN has been assigned to each screw: this preload is calculated, based on a tightening torque $T$ = 106 Nm and oiled surfaces, $\mu_m$ = 0.10 [20–23], as for the company's assembly specification. The displacements retrieved at this step are shown in Figure 12a. A vertical force $F$ = 245.25 N has been added at the second step of the analysis with the bolt preload still acting (lock option set as active in the Ansys WB bolt preload tool). The difference between the displacements recorded at the end of the second step (Figure 12b) and at the end of the first step gives the displacements due to the effect of the external load only, as explained by Equation (2):

$$\Delta y_{FEM} = \Delta y_{step2} - \Delta y_{step1}. \tag{2}$$

Looking at the data in Table 3, relevant to the new FE model, the maximum absolute value of the percentage error is 5% and it is normalized with respect to the displacement at position 2, which is not located on the loaded support but on the adjacent one. Conversely, the errors recorded at the loaded support are of a small entity and stay lower than 3%. In light of the results above, the FE model can now be deemed as validated.

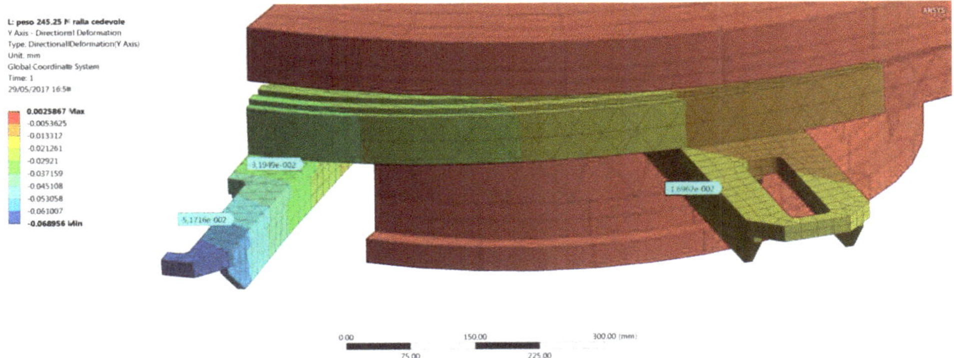

**Figure 10.** FE displacements of the Transfer Machine with nine divisions with $E_{ring}$ = 39 GPa.

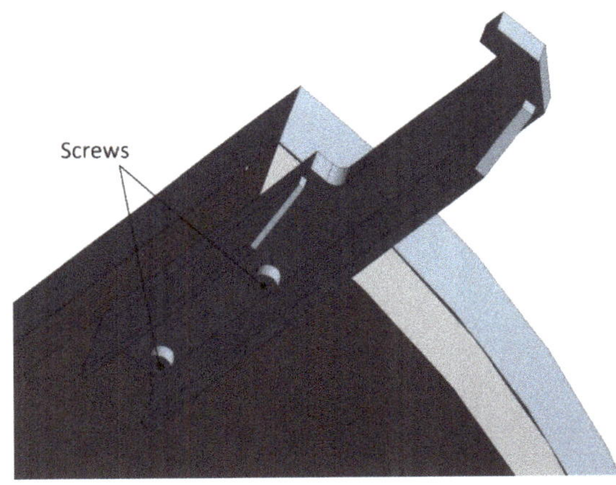


**Figure 11.** Detail of the bolted connection between the rotary table and the support.

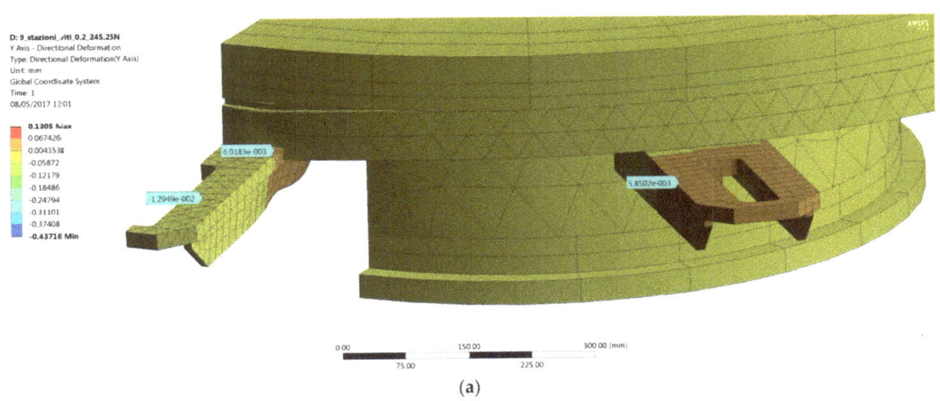

(a)

**Figure 12.** *Cont.*

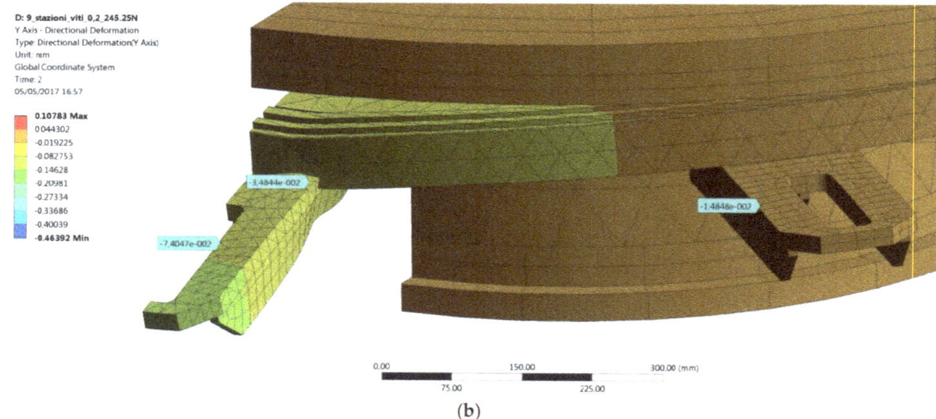

**Figure 12.** (**a**) FE displacements at first step (bolt preload); (**b**) displacements at second step (bolt preload plus external force).

**Table 3.** FE displacements of the modified model at the end of the two steps, $\Delta y_{FEM}$ and errors with respect to the experimental data.

| Position | $\Delta y_{step1}$ (mm) | $\Delta y_{step2}$ (mm) | $\Delta y_{FEM}$ (mm) | $\Delta y_{exp}$ (mm) | Error (-) |
|---|---|---|---|---|---|
| 1 | −0.013 | −0.074 | −0.061 | −0.060 | +1.7% |
| 2 | +0.006 | −0.015 | −0.021 | −0.020 | +5% |
| 3 | +0.006 | −0.035 | −0.041 | −0.040 | +2.5% |

*2.4. Modal Analysis*

Based on the settings reported in the previous sections, considering the bolts as bonded contacts with a reduced stiffness, a modal analysis has been carried out in order to predict the first five modes of vibration of the assembly. The natural frequencies are reported in Table 4 and the first two modes are shown in Figure 13a,b.

The natural frequencies and the associated modes reported above will be useful for a vibration assessment, which should take the following points into account: (i) the match between the deformation induced by the generic excitation force and one of the modes shown above; (ii) the match between the frequency of the excitation force and the frequency of the relevant mode. The source of excitation could either be due to the movement of the carriages or to the cutting operation itself. (iii) A further point is the energy associated to the source of excitation: e.g., it must be carefully assessed if the energy associated to the cutting operation is high enough to bring the structure into resonation, provided that there is compatibility between the modes and frequencies. All these checks have been carried out in a separate study [24].

**Table 4.** FE calculated natural frequencies of the Transfer Machine with nine divisions.

| Mode # | $f$ (Hz) |
|---|---|
| 1 | 80.742 |
| 2 | 81.723 |
| 3 | 89.092 |
| 4 | 117.83 |
| 5 | 143.57 |

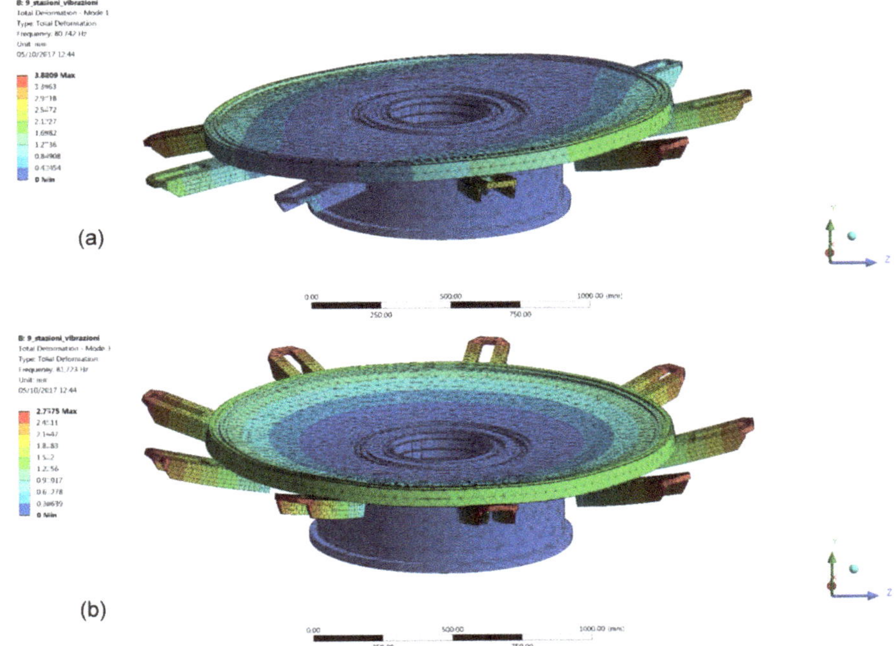

**Figure 13.** Modes of vibration of the Transfer Machine with nine divisions: (**a**) first mode; (**b**) second mode.

## 3. Structural Design of the Transfer Machine with 15 Divisions

### 3.1. FEA and Structural Optimization

The Transfer Machine with 15 divisions is represented in Figure 14: it has six more divisions, thus, if compared to the same machine with nine divisions, a greater diameter ($D_{out}$ = 2940 mm) and one more component (stiffening rib) under the support.

The analysis takes three different load scenarios into account: two radial drilling operations along $x$-axis, positively or negatively oriented, and a vertical ($y$-axis) downwards drilling operation. The purpose of the study is to determine the displacement of the component being clamped under the machining conditions. The same FEM settings validated on the Transfer Machine with nine divisions have been adopted for the analysis of the same machine with 15 divisions, where the same bearing and joining techniques are utilized (see also below). After geometry simplification, the model has been treated again as a symmetrical structure. The materials assigned to the transfer are AlMg0.7Si ($E$ = 69.5 GPa, $v$ = 0.33) for the rotating table and a generic structural steel ($E$ = 200 GPa, $v$ = 0.30) for all other components. The geometry has been meshed with SOLID187 tetrahedral and hexahedral elements, for a total node count $n \approx 160,000$. Every contact between the parts has been set as bonded, except the one between the reference support and the rotating table, which is obtained by means of a set of screws. In particular, the Transfer Machine with 15 divisions has a screw pattern made of six M12 12.9 classes, instead of the four-screw pattern utilized in the same machine with nine divisions, as shown in Figure 15. The modeling strategy is the same adopted for the machine with nine divisions.

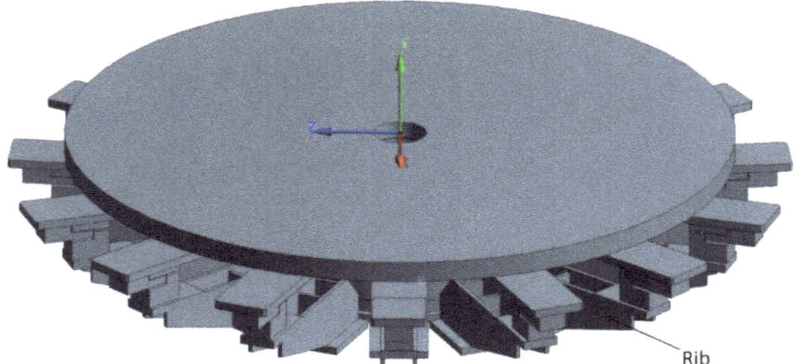



**Figure 14.** Transfer Machine with 15 divisions: a rib under each support helps increase its stiffness.

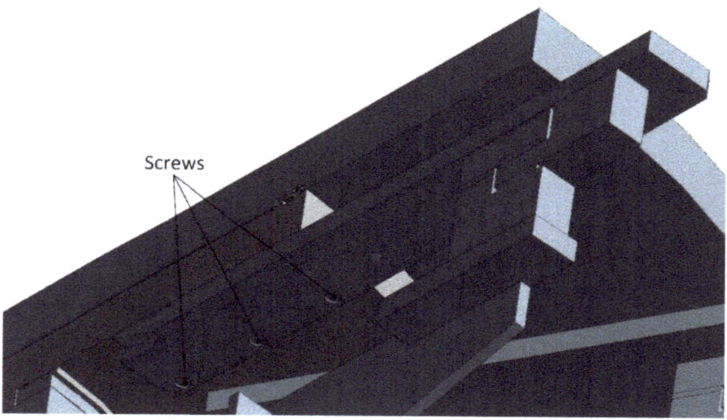



**Figure 15.** Bolted connection between the support and the rotary table of the Transfer Machine with 15 divisions.

Three types of machining scenarios have been studied, so three different simulations had to be performed. Each load has been simulated by a remote force $F_1$ = 500 N acting on the upper surface of a support. Symmetry has been enforced as described above. In order to better approximate the actual area of force application (roughly corresponding to the contact area between the vise and the support) a portion of surface has been isolated by a division line (red area in Figure 16) and the machining force applied thereof. The point of application of the force is the center of gravity of the clamped lock, as shown in Figure 17.

Figure 18 shows the displacements along the $y$-axis due to the sole bolt preload, whereas Figure 19 displays the values of $y$-axis displacement for the three load scenarios. The flagged spot always represents the projection of the center of gravity of the clamped lock on the support: the displacements measured at this point in the three different load cases are summarized in Table 5, along with the calculation of the displacements caused by the sole external loads (the displacement due to the bolt preload has been ruled out by applying Equation (2)).

**Figure 16.** Target surface for the application of remote forces.

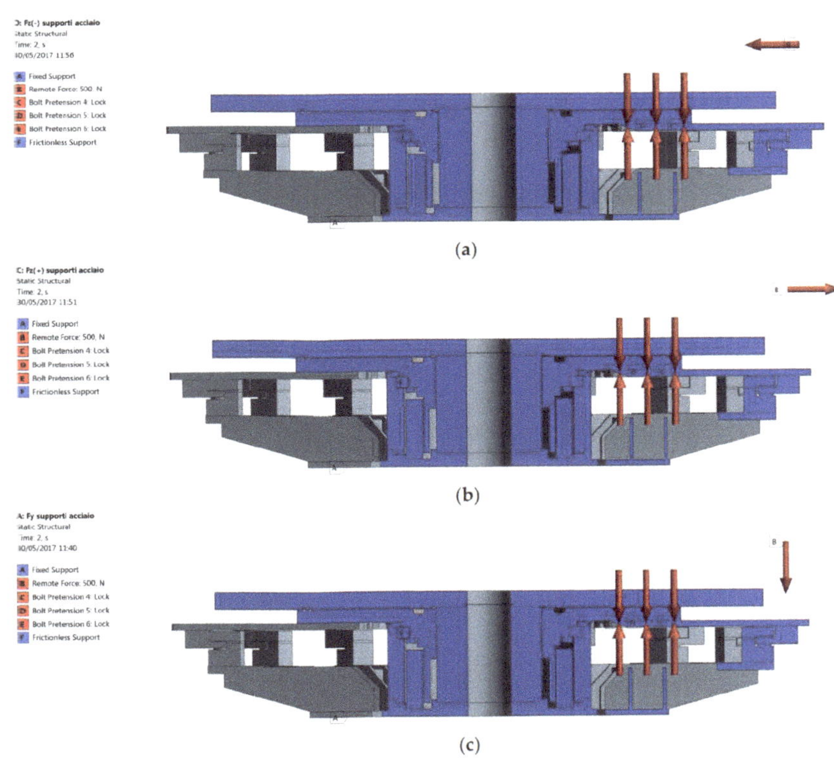

**Figure 17.** Boundary conditions with forces (**a**) $Fz$ $(-)$ (**b**) $Fz$ $(+)$ (**c**) $Fy$ $(-)$. Bolt preload forces.

**Table 5.** FE displacements under different machining conditions for the Transfer Machine with 15 divisions.

| Machining Condition | $\Delta y_{step1}$ (mm) | $\Delta y_{step2}$ (mm) | $\Delta y_{FEM}$ (mm) | $\Delta y_{exp}$ (mm) | Error (-) |
|---|---|---|---|---|---|
| $Fz\ (-)$ | $-0.002$ | $+0.006$ | $+0.008$ | - | - |
| $Fz\ (+)$ | $-0.002$ | $-0.010$ | $-0.008$ | - | - |
| $Fy\ (-)$ | $-0.002$ | $-0.012$ | $-0.010$ | $-0.01$ | 0% |

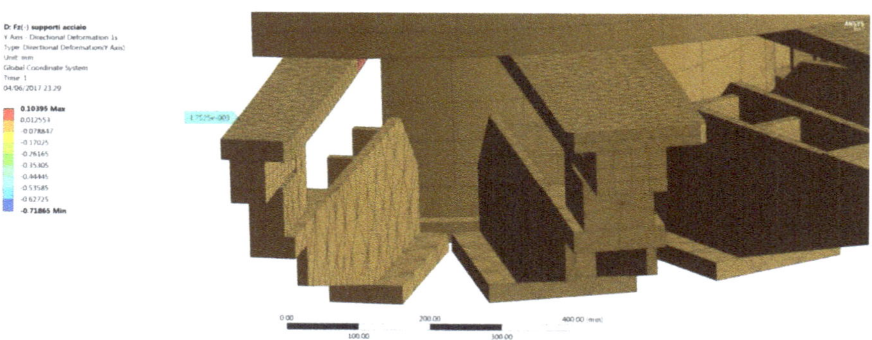

**Figure 18.** FE displacements of the Transfer Machine with 15 divisions after bolt preloading (scale factor 1000×).

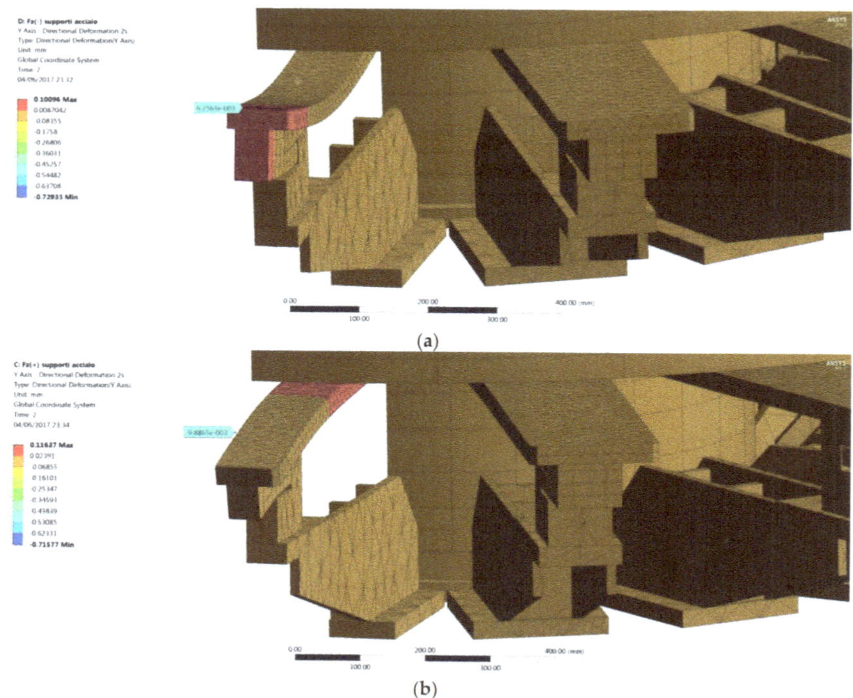

(a)

(b)

**Figure 19.** *Cont.*

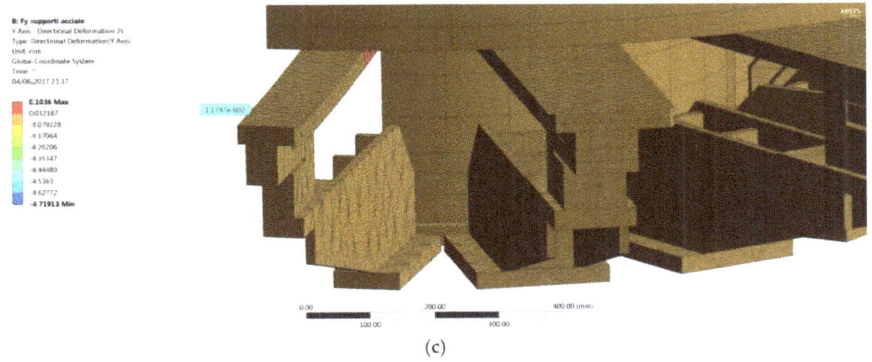

(c)

**Figure 19.** Overall FE displacements of the Transfer Machine with 15 divisions (**a**) $Fz$ (−); (**b**) $Fz$ (+); (**c**) $Fy$ (−).

The greatest displacement on the reference point is caused by the force $Fy$ (−) ($\Delta y = 0.01$ mm), while both $Fz$ (−) and $Fz$ (+) cause a displacement of 0.008 mm. The result reported in Table 5, for the $Fy$ (−) load case only, has been compared with the experimental outcome, retrieved by means of the test setup shown in Figure 20. The FEA outcome perfectly matches the experimental result. It is noteworthy that the vertical displacement of the Transfer Machine with 15 divisions is less than 17% of that for the same machine with nine divisions, under the same loading condition of vertical downwards drilling, even though the first has a much larger (and thus more flexible) rotary table. Such an outcome has been achieved thanks to a dynamic stiffening system developed by the authors. In fact, the stiffening ribs do not spin around the $y$-axis together with the rotary table; they are instead fixed to the ground. The ribs and the supports at the outer diameter of the rotary table (Figure 21a: in blue shades the moving parts) are clamped together just for the machining time, by means of a hydraulic clamp whose operating principle is quite the same of a disc brake (Figure 21b). This simple device allows for keeping the rotary table comparatively light, as well as achieving enough bending stiffness when needed.

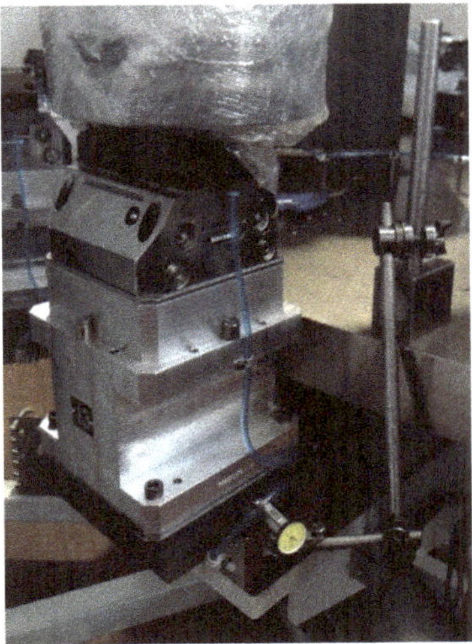

**Figure 20.** Experimental measurement of the vertical displacement of the support in the case of $Fy$ (−) = 500 N.

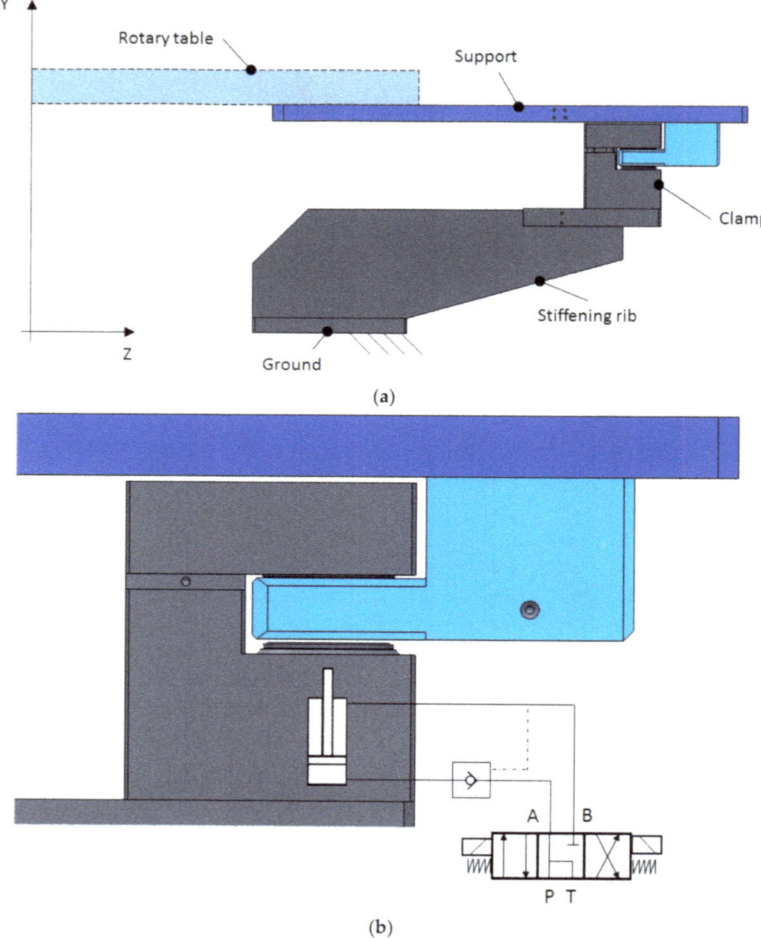



**Figure 21.** Hydraulic locking clamp (**a**) nomenclature and geometry and (**b**) operating principle and hydraulic schematic (P—pressure line, T—tank line).

In order to assess the possibility of further weight savings on the moving parts, further FEA have been run by changing the material assigned to the supports: e.g., lightweight alloys: aluminum, titanium, and magnesium. The comparison between the different solutions has been made in the case of $Fy$ (−). The vertical displacements have been sampled along the path illustrated in Figure 22. The results in terms of vertical displacement for different choices of the support material, normalized with respect to the steel support are shown in the plot of Figure 23, whereas Figure 24 shows the displacements obtained with different materials combinations of the support and of the rotary table, normalized with respect to the steel-steel combination.

**Figure 22.** Sampling path defined along an edge of the loaded support.

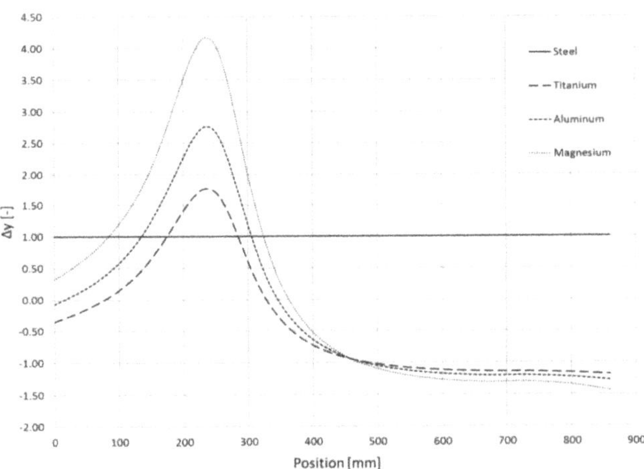

**Figure 23.** FE displacements of the loaded support (normalized with respect to the steel support) with $Fy$ ($-$) force: comparison between different materials of the support.

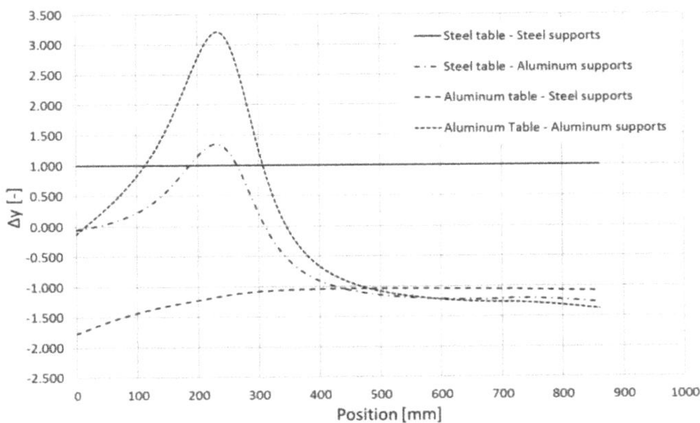

**Figure 24.** FE displacements of the loaded support (normalized with respect to the steel-steel combination.) with $Fy$ ($-$) force: comparison between different material combinations of the rotary table and the supports.

As can be appreciated by looking at Figure 23, besides Ti-alloys, which would have been impractical due to cost and manufacturing reasons, both Al-alloy and Mg-alloy supports negatively affect the bending stiffness of the assembly, especially at the outer edge where machining operations take place: a steel support is therefore the best option. On the other hand, looking at Figure 24, it can be seen that a combination of Al-alloy table and steel support would not be much more compliant than the base case with both components made of steel. The displacement at the outer end would increase by just $10^{-3}$ mm, should the lighter construction be adopted.

Based on these premises, an aluminum alloy has been chosen for the construction of the rotary table, whereas the supports have been built of steel.

### 3.2. Modal Analysis

Based on the settings reported in the previous sections for the Transfer Machine with nine divisions, a modal analysis has been carried out in order to determine the first five modes of vibration of the same machine with 15 divisions. The natural frequencies are reported in Table 6 and the first two modes are shown in Figure 25a,b.

**Table 6.** FE calculated natural frequencies of the Transfer Machine with 15 divisions.

| Mode # | $f$ (Hz) |
|--------|----------|
| 1 | 123.74 |
| 2 | 147.61 |
| 3 | 154.31 |
| 4 | 154.97 |
| 5 | 155.01 |

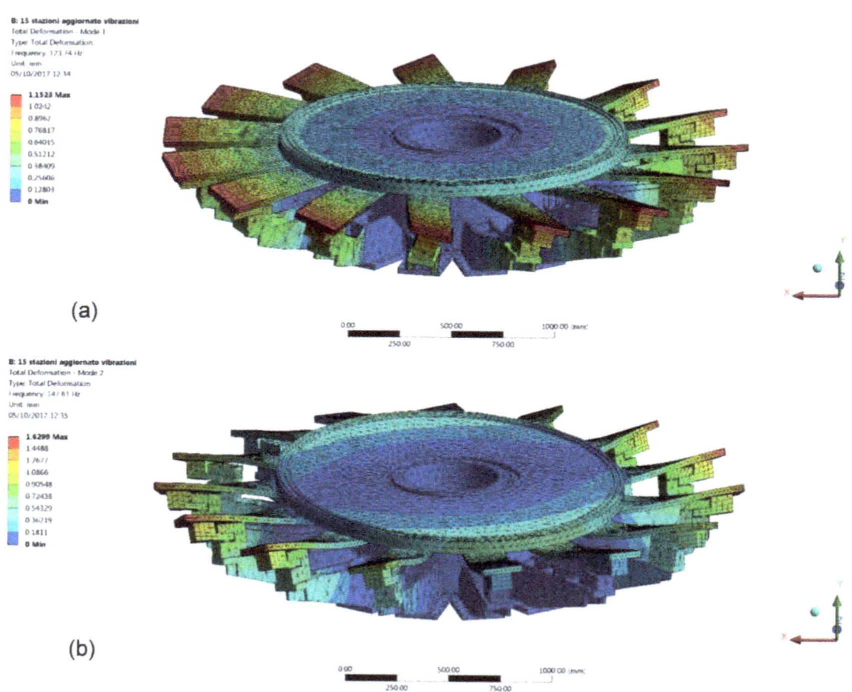

(a)

(b)

**Figure 25.** Modes of vibration of the Transfer Machine with 15 divisions: (**a**) first mode; (**b**) second mode.

## 4. Discussion

In light of the results presented in the previous sections, some general guidelines for the structural design of a modern transfer machine tool can be drawn.

1.  Different FE modeling strategies lead to different predictions in terms of the overall stiffness of the assembly. The following considerations stem from the results illustrated at Section 2.3: (i) the accurate modeling of the bearing stiffness has little or no impact on the stiffness prediction provided by FEA. Therefore, this component can be even modeled as a single body with the elastic properties of steel without significantly affecting the accuracy of the results; (ii) the accurate modeling of the bolt pattern has conversely a significant effect on the stiffness of the FE model. As a rule of thumb, at preliminary analysis stage, or when the bolt preload tool cannot be used, a bonded contact with a pure penalty formulation and a normal stiffness factor set at $FKN = 1.5 \times 10^{-3}$ has been proven to provide results close to the case of bolt modeling. The FKN parameter has to be fine-tuned based on the actual application.

2.  Other rather complex components, such as roller monoguides, ballscrews and the like can be represented with a good accuracy in terms of stiffness, by replacing them with homogeneous, isotropic elastic single bodies, whose elastic modulus and Poisson's ratio shall be tuned by means of preliminary FEA. However, it must be observed that the loading scheme adopted in the tuning FEA must be properly chosen in order to properly reproduce the actual load scenario in the machine service conditions.

3.  The bigger, 15-division transfer, needs some stiffness raisers, so as to match the performance of the same machine with nine divisions. For instance, a rib, which is fixed to the ground, helps increase the flexural stiffness of the rotary table–supports assembly. The rib stiffness kicks in just during the machining time, by means of a hydraulic clamp device developed by the authors. Such an arrangement allows for achieving a twofold task: (i) proper stiffness is warranted during machining; (ii) the moving parts are kept as light as possible, thus shrinking the chip-to-chip time. The advantage in terms of stiffness may be appreciated considering that, despite its larger size, the maximum displacement of the transfer machine with 15 divisions is approximately one-fourth of the displacement of the corresponding machine with nine divisions for given vertical load, measured at the same reference point.

4.  With regard to the 15-division transfer, some FEAs have been performed in order to work out the most favorable combination between the materials for the construction of the supports and of the rotary table. This combination turned out to be the one comprising an aluminum alloy table and steel supports. No significant loss in terms of flexural stiffness has been observed with respect to the base case with all steel components.

5.  Based on the outcomes of the modal analysis, should any loading condition lead to instability, this issue could be overcome and the system optimized by locally modifying the stiffness of the elements and/or the distribution of the masses.

## 5. Conclusions

An experimentally validated numerical model for transfer machine tools has been defined, which allows for forecasting the structural as well as the vibrational response of this kind of machine under typical operating loads. The models helped develop a bigger machine, evaluating different scenarios in terms of materials and design solutions. Through this approach, it has also been possible to optimize the structure in terms of flexural stiffness at a very early stage of the design. A proper choice of the materials and the introduction of an ad hoc designed stiffening solution made it possible to achieve even greater stiffness than with the smaller machine, thus ensuring the desired manufacturing tolerance of the finished parts. This choice will ultimately lead to higher speed and a more efficient machine whose performance is aligned with the values of green design.

**Author Contributions:** D.C. and N.V. conceived and designed the experiments; O.C. performed the experiments; M.D.A., S.F., and F.R. performed the numerical analyses; G.O. analyzed the data; N.V. provided reagents, materials and analysis tools; O.C., S.F. and M.D.A. wrote the paper.

**Conflicts of Interest:** The authors declare no conflict of interest.

## List of Symbols

| | | |
|---|---|---|
| $F$ | Spindle thrust force | (N) |
| $m$ | Mass | (kg) |
| $n$ | Number of total nodes (FEA) | (-) |
| $\Delta y$ | Displacement along $y$-axis | (mm) |
| $E$ | Young's modulus | (GPa) |
| $\nu$ | Poisson's ratio | (-) |
| $\delta$ | Assembly preload displacement of the bearing | ($\mu$m) |
| $E_{ring}$ | Equivalent elastic modulus of the bearing | (GPa) |
| $\mu$ | Friction coefficient | (-) |
| $\mu_m$ | Mean friction coefficient of a bolted joint | (-) |
| $F_i$ | Bolt preload | (N) |
| $T$ | Tightening torque | (Nm) |
| $D_{out}$ | Outer diameter of the table | (mm) |

## References


1. European Commission. Commission Regulation (EC) No 641/2009 of 22 July 2009. *Off. J. Eur. Union* **2009**, *L191*, 35–41.
2. Pervaiz, S.; Deiab, I.; Rashid, A.; Nicolescu, M. Minimal quantity cooling lubrication in turning of Ti6Al4V: Influence on surface roughness, cutting force and tool wear. *Proc. Inst. Mech. Eng. Part B J. Eng. Manuf.* **2017**, *231*, 1542–1558. [CrossRef]
3. Gupta, M.K.; Sood, P.K. Machining comparison of aerospace materials considering minimum quantity cutting fluid: A clean and green approach. *Proc. Inst. Mech. Eng. Part C J. Mech. Eng. Sci.* **2017**, *231*, 1445–1464. [CrossRef]
4. Brockhoff, T.; Walter, A. Fluid minimization in cutting and grinding. *Abrasives* **1998**, *10*, 38–42.
5. Sharma, V.S.; Singh, G.; Sørby, K. A review on minimum quantity lubrication for machining processes. *Mater. Manuf. Process.* **2015**, *30*, 935–953. [CrossRef]
6. Pejryd, L.; Beno, T.; Isaksson, M. Machining aerospace materials with room-temperature and cooled minimal-quantity cutting fluids. *Proc. Inst. Mech. Eng. Part B J. Eng. Manuf.* **2011**, *225*, 74–86. [CrossRef]
7. Chatha, S.S.; Pal, A.; Singh, T. Performance evaluation of aluminium 6063 drilling under the influence of nanofluid minimum quantity lubrication. *J. Clean. Prod.* **2016**, *137*, 537–545. [CrossRef]
8. Saikawa, Y.; Ichikawa, T.; Aoyama, T.; Takada, T. High speed drilling and tapping using the technique of spindle through MQL supply. *Key Eng. Mater.* **2004**, *257–258*, 559–564. [CrossRef]
9. Treurnicht, N.F.; Joubert, H.J.; Oosthuizen, G.A.; Akdogan, G. Investigating of eco- and energy-efficient lubrication strategies for the drilling of light metal alloys. *S. Afr. J. Ind. Eng.* **2010**, *21*, 25–38. [CrossRef]
10. Itoigawa, F.; Childs, T.; Nakamura, T.; Belluco, W. Effects and mechanisms in minimal quantity lubrication machining of an aluminum alloy. *Wear* **2006**, *260*, 339–344. [CrossRef]
11. Braga, D.U.; Dinizc, A.; Miranda, G.; Coppini, N. Using a minimum quantity of lubricant (MQL) and a diamond coated tool in the drilling of aluminum–silicon alloys. *J. Mater. Proc. Technol.* **2002**, *122*, 127–138. [CrossRef]
12. Qin, S.; Li, Z.; Guo, G.; An, Q.; Chen, M.; Ming, W. Analysis of minimum quantity lubrication (MQL) for different coating tools during turning of TC11 titanium alloy. *Materials* **2016**, *9*, 804. [CrossRef] [PubMed]
13. Azarrang, S.; Baseri, H. Selection of dry drilling parameters for minimal burr size and desired drilling quality. *Proc. Inst. Mech. Eng. Part E J. Process Mech. Eng.* **2017**, *231*, 480–489. [CrossRef]
14. *Theory Reference for ANSYS and ANSYS Workbench, Release 11*; SAS IP Inc. Southpointe, 275 Technology Drive: Canonsburg, PA, USA, 2007.

15. Croccolo, D.; De Agostinis, M.; Fini, S.; Morri, A.; Olmi, G. Analysis of the influence of fretting on the fatigue life of interference fitted joints. In Proceedings of the ASME International Mechanical Engineering Congress and Exposition, (IMECE), Montreal, QC, Canada, 14–20 November 2014.
16. Croccolo, D.; De Agostinis, M.; Vincenzi, N. Structural analysis of an articulated urban bus chassis via FEM: A methodology applied to a case study. *Stroj. Vestnik J. Mech. Eng.* **2011**, *57*, 799–809. [CrossRef]
17. Xu, Z.; Xi, F.; Liu, L.; Chen, L. A method for design of modular reconfigurable machine tools. *Machines* **2017**, *5*, 5. [CrossRef]
18. Croccolo, D.; De Agostinis, M.; Fini, S.; Olmi, G. A user-friendly computational algorithm for the structural analysis of wrapping machine rotating rings. *Proc. Inst. Mech. Eng. Part C J. Mech. Eng. Sci.* **2016**, *230*, 2776–2791. [CrossRef]
19. Croccolo, D.; De Agostinis, M.; Fini, S.; Olmi, G. Analysis of threaded connections for differential gear pinions. In Proceedings of the ASME International Mechanical Engineering Congress and Exposition, (IMECE), Phoenix, AZ, USA, 11–17 November 2016.
20. De Agostinis, M.; Fini, S.; Olmi, G. The influence of lubrication on the frictional characteristics of threaded joints for planetary gearboxes. *Proc. Inst. Mech. Eng. Part C J. Mech. Eng. Sci.* **2016**, *230*, 2553–2563. [CrossRef]
21. Croccolo, D.; De Agostinis, M.; Fini, S.; Olmi, G. Tribological properties of bolts depending on different screw coatings and lubrications: An experimental study. *Tribol. Int.* **2017**, *107*, 199–205. [CrossRef]
22. Zhao, C.; Zheng, W.; Ma, J.; Zhao, Y. Shear strengths of different bolt connectors on the large span of aluminium alloy honeycomb sandwich structure. *Appl. Sci.* **2017**, *7*, 450. [CrossRef]
23. Fallahnezhad, K.; Steele, A.; Oskouei, R.H. Failure mode analysis of aluminium alloy 2024-T3 in double-lap bolted joints with single and double fasteners; A numerical and experimental study. *Materials* **2015**, *8*, 3195–3209. [CrossRef]
24. Martini, A.; Troncossi, M.; Rivola, A.; Vincenzi, N. Experimental vibration analysis of a rotary transfer machine for the manufacture of lock components. In Proceedings of the Surveillance 9 International Conference, Fes, Morocco, 22–24 May 2017; pp. 1–9.

MDPI
St. Alban-Anlage 66
4052 Basel
Switzerland
Tel. +41 61 683 77 34
Fax +41 61 302 89 18
www.mdpi.com

*Machines* Editorial Office
E-mail: machines@mdpi.com
www.mdpi.com/journal/machines

www.ingramcontent.com/pod-product-compliance
Lightning Source LLC
LaVergne TN
LVHW070652170726
843515LV00004B/387